ISBN 978-1-332-17944-2
PIBN 10294483

POCKET

Ophthalmic Dictionary

INCLUDING

Pronunciation, Derivation and Definition

OF THE WORDS

USED IN

Optometry and Ophthalmology

Together with a complete description of the light wave theory, Anatomy of the Eye, functions, blood and nerve supply of the different parts, Retinoscope, Ophthalmoscope, Trial Case and how to use them. Rules, Transposition, Toric and other lenses. State board questions.

By JAMES J. LEWIS, Oph. D.

Professor of Optometry in the Northern Illinois College of Ophthalmology and Otology, Chicago.

FIFTH EDITION

Revised and Enlarged **Illustrated**

To the Profession:

This edition has been compiled with a great deal of care. Realizing that perfection in its full sense has never been attained by mortal man, the author invites the unbiased and conscientious criticism of the readers and users of this Dictionary and hereby earnestly solicits the same to the end that the future editions may profit by the honest convictions of studious oculists, physicians and optometrists.

I cannot conclude without expressing my deep sense of obligation to the profession for their kind reception of this work. Feeling the responsibility incurred by those who attempt to teach others, I have spared no amount of labor or cost to render this volume clear, practical and useful.

Very respectfully,

THE AUTHOR.

PREFACE

TO THE FIFTH EDITION

THE very favorable reception accorded the fourth edition of this work has encouraged the author to still further revise it, incorporating in its pages a considerable number of new definitions, as well as giving an accurate and complete derivation of all the technical terms used in Optical nomenclature, making the book of the highest value to the practitioner and student alike.

In presenting this fifth edition of the Lewis Dictionary and Encyclopedia, I wish to express my appreciation for the invaluable assistance extended by

J. B. McFATRICH, M. S., M. D.,

Professor of the Principles of Ophthalmology and Otology.

GEO. WILBUR McFATRICH, M. D.,

Professor of Clinical and Didactic Ophthalmology and Otology.

HENRY S. TUCKER, A. M., M. D.,

Professor of Anatomy and Physiology of the Eye and Brain.

E. G. TROWBRIDGE, M. D.,

Professor of Dioptrics.

J. J. L.

660

Abbreviations and Optical Signs

Acc.......................Accommodation.
Aet.......................Age.
Am.......................Ametropia.
An.......................Anisometropia.
As.......................Astigmatism.
Asth.....................Asthenopia.
Ax.......................Axis.
Cc. or — (minus).........Concave.
Ce.......................Centigrade.
Cm.......................Centimeter.
Cx. or + (plus)..........Convex.
Cyl......................Cylinder.
D........................Dioptry.
D. Cc....................Double concave.
D. Cx....................Double convex.
D. T.....................Distance test.
E. or Em.................Emmetropia.
e. g.....................For example.
H. or Hy.................Hypermetropia.
In.......................Inches.
L. or L. E...............Left eye.
M. or My.................Myopia.
Mm.......................Millimeter.
N........................Nasal.
Nv.......................Naked vision.
O. D. (Oculus Dexter)....Right eye.
O. S. (Oculus Sinister)...Left eye.
O. U. (Oculi Unati)......Both eyes.
P. or Pb.................Presbyopia.
P. Cc....................Periscopic concave.
P. Cx....................Periscopic convex.

Abbreviations and Optical Signs—Con.

P. D	Inter-Pupillary distance.
Pl.	Plano.
p. p. (Punctum Proximum)	Near point.
p. r. (Punctum Remotum)	Far point.
Pr.	Prism.
R. or R. E	Right eye.
R. T.	Reading test.
Rx	Prescription.
Sb.	Strabismus.
S. or Sph.	Spherical.
T.	Temporal.
Ty	Type.
V.	Vision.
Va	Visual acuteness.
W. P.	Working point.
+	Plus convex.
—	Minus — concave.
◯	Combined with.
	Degree.
△	Prism Dioptry.
=	Equal to.
∞	Infinity, 20 ft. or farther.
′	Line, 12th part of inch.
±	Plus or (and) minus.
▽	Centrad.
×	Multiplied by; times.
÷	Divided by.
>	Is (or are) greater than.
<	Is (or are) less than.

Lewis Ophthalmic Dictionary

A (Gr. alpha, privative.) A prefix conveying a negative meaning; as without, away, not, from.

Abaxial (ab-aks'-e-al). Not situated in the line of the axis.

Abbe, Prof. Ernst. A German professor and inventor of an apparatus for measuring the indices of refraction. 1845-1905.

Abducens (ab-du'-senz). (L. ab. = from + ducere = to draw.) The external rectus muscle, whose action it is to rotate the eye outward. It arises by two heads, one from the lower margin of the sphenoidal fissure; the other from the outer margin of the optic foramen. Its tendon is inserted into the sclera 8 mm. from the outer margin of the cornea. Under normal conditions these muscles should overcome about 8 degrees of prism, base in. It is controlled by the sixth pair of cranial nerves (abducens).

Abducent (ab-du'-sent). Abducting; drawing from the center.

Abduct (ab = away, ducere = to lead). To draw away from the median line.

Abduction (ab-duc'-shun). The act of turning the eye outward from its position of rest. For testing the power of the abductors or external recti muscles, use the strongest prism, base in, with which the eyes can overcome diplopia, looking 20 feet away, and deduct any imbalance.

Abductor (ab-duc′-tor). Any muscle that abducts. For instance, abductor oculi, the external rectus muscle.

Aberration (ab-er-a′-shun). (L. ab = away, errare = to wander.) Wandering from normal. When applied to lenses would mean, unable to obtain a perfect focus. It is due to the greater refractive power of the edge over the center of convex lenses, thus causing the image to be somewhat blurred. In the eye the iris shuts off the edge of the lens, and in this way prevents spherical aberration. **Chromatic Aberration,** dispersion of colors. Owing to the colored rays having different degrees of refractibility they are not focused at the same distance.

Ablatio-retinae (ab-la′-she-o-ret′-in-e). (L. ab = away + latum = to take.) Detachment of the retina.

Ablepharia (ah-blef-ar′-e-ah). (Gr. a = not + blepharon = eyelid.) That condition in which the eyelids are absent. Also, **Ablepharous.**

Ablepsia (ah-blep′-se-ah). (Gr. a = not + blepo = I see.) Blindness—want of sight.

Abnormal (L. ab = away + normal = rule). Away from normal. Relating to vision would mean, any defect of sight. **(Ametropia.)** An eye wherein parallel rays of light do not focus on the retina with the muscles of accommodation at rest.

Abrasio-cornea (ab-ra′-si-o-cor′-ne-ah). (L. ab = away + radere = to scrape.) The rubbing off of the outer layer of the cornea.

Abscess (ab′-ses). (L. ab = away + cedere = to

depart.) A collection of pus in any cavity formed by the separation of tissue.

Abscissa (ab-sis′-a). (L. cut off.) A certain line used in determining the position of a point in a plane.

Absorption (ab-sorp′-shun). (L. ab, and sorbere = to suck in.) A term applied in the operation for cataract where the lens capsule is needled, allowing the aqueous humor to absorb the lens.

Absorptive. Anything that has the power of absorption.

Abstract (ab′-strakt). (L. abstractus = drawn away.) An abstract number is a number not designated as referring to any particular class of objects.

Acceleration. An increase in rapidity; opposed to retardation.

Accommodation (ak-kom′-mo-da′-shun). (L. accommodare = to depart.) The act of adjusting the eye to see within its far point of vision. Optical adjustment. It takes place by contracting the ciliary muscles which encircle the crystalline lens and draws forward the inner layer of the choroid and hyaloid membrane, the suspensory ligaments becoming relaxed, and the lens (by its own elasticity) allowed to assume a greater convexity, especially its anterior surface, thus increasing its refraction. It is never used unless the light attempts to focus behind the retina. **Amplitude of Accommodation** is the difference in the dioptric power of the eye when in a state of complete relaxation and when the full amount of accommodation is in use; or, in other words, the amount of accommodation an eye possesses. Amplitude of Accommodation at different ages (from Landolt) as follows:

Age in Years	Amplitude (dioptries)
10	14
15	12
20	10
25	8.5
30	7.0
35	5.5
40	4.5
45	3.5
50	2.5
55	1.75
60	1.0
65	0.75
70	0.0

This is approximately correct, but individuals differ in the amount of accommodation they possess at the same age. Accommodation is spoken of as binocular, absolute, and relative.

Binocular Accommodation is the full amount of accommodation which both eyes can use together while converging.

Absolute Accommodation is the total amount of accommodation of one eye only, the other being covered.

Relative Accommodation is the amount of accommodation that can be used without changing the convergence; that is, by lenses or other means.

Spasm of Accommodation, the inability on the part of the patient to relax his accommodation without drugs.

Center of Accommodation is situated beneath the floor of the aqueduct of Silvius.

Paralysis of Accommodation is the loss of power of movement in the ciliary muscles through injury, disease, or a drug affecting the nerve supply.

Negative Accommodation would mean that the eye possesses the ability to decrease its dioptric power from that which it possesses in a state of rest. By so doing would lengthen its principal focus.

Accommodation is interfered with by hardening of the lens, weakness of the ciliary muscles, paralysis of the third nerve, loss of the crystalline lens, or use of drugs, such as Atropine.

Achloropsia (Gr. a = without + chloros = green + opsis = vision). Green-blindness, color blindness as regards green.

Achroma (ak-ro′-mah). (Gr. a = not + chroma = color.) Without color.

Achromatic Lens (ah-kro-mat′-ik). (See Lens.)

Achromatism (ah-kro′-ma-tism). Absence of chromatic aberration.

Achromatistous (ah-kro-mat-is′-tus). Deficient in coloring matter or pigment.

Achromatopsia (ah-kro-mat-op′-se-ah). (Gr. a = lacking, chroma = color, eye.) Total color-blindness.

Achromatosis (ah-kro-mat-o′-sis). Any disease marked by lack of pigmentation.

Acorea (ah-ko′-re-ah). (Gr. a = not + kore = pupil.) When the pupil is absent.

Acquired. Not born with, but developed after birth.

Acuity (ak-u'-it-e). (L. acuere to sharpen.) Sharpness, like a needle. The sharpness of vision; the keenness of the visual powers. The acuteness of vision means the vision the patient has with his full correction. The faculty of the retina to perceive forms depends on many conditions:

1. Primarily, on the sensibility of the retina.
2. On the adaptation of the retina.
3. On the general illumination.
4. On the sharpness of the retinal image.
5. On the intensity of the illumination.

It is known that the acuteness of vision varies with the general illumination up to a certain degree of intensity, as that of a clear, sunny day; the two then vary in a direct proportion, but when the illumination passes a certain limit of intensity, the acuteness of vision diminishes instead of increases.

Adaptation (ad-ap-ta'-shun). (L. adaptare to adapt.) Adjustment of the pupil to light.

Addition (a-dish'-on). (L. addere to increase.) The uniting of two or more numbers in one sum.

Adducens (ad-du'-sens). (L. ad towards, ducere to lead.) When this term is applied to the eye it means the internal rectus muscle, the muscle which turns the eyeball inward toward the nose, supplied by the third cranial or motor oculi nerve. The power of adduction of the eye ranges from 20 up to 50 degrees. For testing the power of the adducens or internal rectus muscle, use the strongest prism, base out, with which the eyes can overcome diplopia.

Adduct (L. adducere = to bring toward). To draw inward toward a center.

Adduction. Movement of the eyeball inward from its position of rest. The adducens means the internal rectus muscle by which we turn the eyes inward. The test for the power of the adducens is made by first correcting any error of refraction and have the patient look at a light 20 feet away, placing the base out of the strongest prism, with which the eyes can overcome diplopia. We figure from their position of rest and this prism registers the adduction.

Adenectomy (ad-en-ek′-tō-me). (Gr. aden = gland + ektome = excision.) Removal of a gland by operation.

Adenemphraxis (Gr. aden = gland + emphraxis = stoppage). That condition in which the duct or gland is obstructed.

Adenoid (ad′-en-oid). (Gr. aden = gland + edois = appearance.) Resembling a gland.

Adenologaditis (ad′-en-o-log-ad-i-tis′). Inflammation of the glands of the eyes and conjunctiva. Ophthalmia neonatorum.

Adenophthalmia (ad-en-off-thal′-me-ah). (Gr. aden = gland + ophthalmos = eye.) Inflammation of the meibomian glands.

Aditus (ad′-i-tis). (L. "a passage.") The entrance to a canal or duct. **Aditus Orbitae,** the opening of the orbit, covered by the eyelids.

Advancement. The cutting away of a muscle of the eye and attaching it to an advanced point.

This operation is performed on the weak muscle in cases of strabismus.

Adventitious (ad-ven-tish'-us). Acquired—not normal.

Afferent (L. ad = to + ferre = to bear). Conveying from the surface to the center, as a nerve or vein.

Albinism (al'-bin-ism). (L. albus = white.) Abnormal deficiency of pigment in the iris and choroid.

Albugo (al-bu'-go). White opacity of the cornea of the eye. Leukoma.

Alexia (a-lex'-ia). (Gr. a = lacking, lexia = word.) Unable to read, due to a central lesion.

Amaurosis (am-aw-ro'-sis). (Gr. amaurotin = to render obscure.) A disease of the optic nerve or retina, which causes blindness.

Ambiopia (am-be-o'-pe-ah). (Gr. ambo = both, opia = eye.) Vision with both eyes.

Amblyopia (am-ble-o'-pe-ah). (Gr. amblys = blunt + opsis = sight.) A dimness of vision from defective sensibility of the retina. A condition in which there is a possibility of restoring the former vision; for instance, when a person has an error of refraction in one eye, the other eye being emmetropic, he will learn to ignore the eye with the error, and use the one with the best vision. In this way the sight will become dim from want of use, and is an acquired state, which by testing with the pinhole disc will show no improvement. Under these conditions, the error must be corrected with the retinoscope, and if the eyes are not more than two dioptries apart, instruct your patient always to wear his correction and cover the good eye two or three times a day, for a period of ten minutes at a time, and try to use the amblyopic eye. In this way you will notice an improvement of vision each week. When the pinhole disc fails to improve vision, the eye is either amblyopic or in a diseased state. **Toxic Amblyopia** is a dimness of vision from the poisonous effect of drugs, such as quinine, upon the nervous system; excessive use of tobacco or alcoholic stimulants produce the same effect. The treatment for this form of Amblyopia does not consist of glasses, but the patient must quit the use of the drug causing the trouble, and if not too far advanced there is a possibility of recovering the former vision. **A., Postmar'ital,** that due to sexual excess. **A., Crossed,** on one-half of retina.

Amblyopia ex Anopsia. Amblyopia resulting from one eye having been excluded for some time from binocular vision.

Ametrometer (a-met-rom'-e-ter). (Gr. a = lacking + metron = measure.) An instrument used for measuring ametropia.

Ametropia (a-met-ro'-pe-ah). (Gr. a = lacking, metron = measure, ops = eye.) Any error of refraction, such as hyperopia, myopia, or astigmatism. **Axial Ametropia.** Ametropia that is caused by the length of the eyeball on the optic axis. The opposite of Emmetropia.

Amphice'lous (Gr. amphi = on both sides + koilos = hollow). Concave on both sides or ends.

Amphodiplopia (am-fo-dip-lo'-pe-ah). (Gr. ampho

= both + diploos = double.) That conditio where both eyes have double vision.

Amplifier (am′-ple-fi-er). An apparatus for in creasing the magnifying power of a microscope

Amplitude (am′-pli-tud). State of being ample (Physics) The extent of a movement measure from the starting point or position of rest. (Math.) An angle upon which the value of some function depends.

Amplitude of Convergence. (See Convergence.)

Amyosta′sia (Gr. a = not + mys = muscle + stasis = standing). Nervous tremor of the muscles.

Amyosthe′nia (Gr. a = not + mys = muscle + sthenos = strength). Failure of muscular strength.

Anacamptom′eter. An instrument for measuring the reflexes.

Anaclasis (an-ak′-las-is). (Gr. bending back, reflection.) When this term is applied to light, it refers to the rays traveling obliquely from a rarer to a denser medium, being bent backward toward the perpendicular (refraction).

Anaesthesia (an-es-the′-ze-ah). (Gr. an = not + aisthesis = sensation.) Lacking sensitiveness, where the retina is amblyopic.

Anaesthetic (an-es-thet′-ik). The name given to that which produces insensibility to pain, as chloroform, cocaine, and ether.

Analysis (a-nal′-i-sis). (Gr. "a releasing.") A resolution of a whole into its parts; a form of reasoning from a whole to its parts.

Anaphoria (an-a-phor′-ia). (Gr. ana = up + phoria = tending.) That condition in which the eyes turn upward when the extrinsic muscles are in a state of rest. Stevens gives 33 degrees for the maximum elevation.

Anastomosis (a-nas-to-mo′-sis). (Gr. anastomoo = "through a mouth.") The junction of vessels; the joining of blood-vessels with one another by means of branches, whereby, if the fluid be arrested in its course through one vessel, it will proceed through others. The term is sometimes applied to the junction of nerve filaments with each other.

Anerythrop′sia. Red-blindness.

Antecedent (an-te-se′-dent). (L. antecedere = to go before.) The first of two terms of a ratio.

Anatomy (a-nat′-o-me). (Gr. anatome = dissection; ana = up, tome = a cutting.) **(Eye.)** Relates to the description of the structures of the eye and its parts. The eyeball is nearly spherical in shape and measures about 24 mm. in diameter. The cornea represents a segment of a small sphere projecting from its anterior surface. The first tunic of the eyeball is the sclerotic and cornea. The posterior five-sixths is the sclerotic, which is white and opaque, and serves to give shape to the eye and protects its more delicate interior. Near the posterior pole, on the nasal side, is a sieve-like disc known as the lamina cribrosa, through which the optic nerve fibers enter the eye. The sclerotic is thickest at its posterior portion and gradually becomes thinner as it approaches the equator, and again thickens as it approaches the cornea. The anterior one-sixth is the cornea. It is transparent and of a greater curvature than the sclerotic. The cornea is set in the sclerotic as a watch crystal is placed in its frame, and is composed of five layers, from without inward as follows: Conjunctiva Epithelium, Bowman's Membrane, Cornea Proper, Membrane of Descemet, and the Endothelium. At the inner angle (angle of filtration) between the iris and cornea, there are a number of comb-like openings which are in the trabecular tissue or pectinate

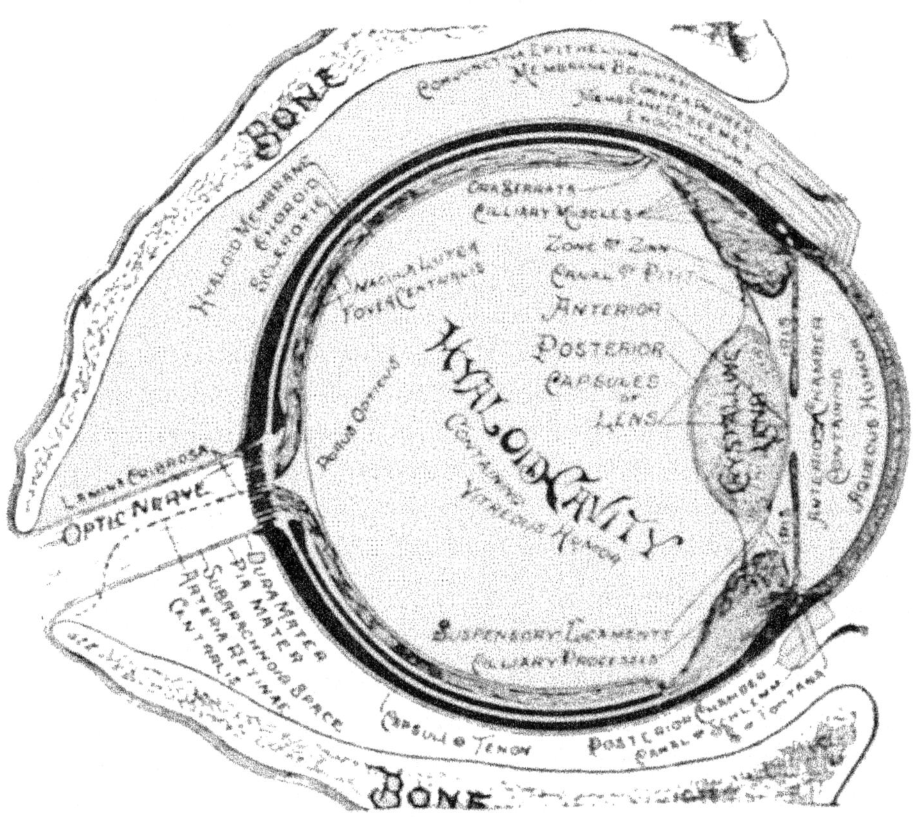

ligament which runs from the periphery of the cornea to the base of the iris. These openings are called the spaces of Fontana, through which the aqueous humor passes into the canal of Schlemm, a circular canal extending around the periphery of the cornea at the sclero corneal junction. From this canal the humor passes into the anterior ciliary veins. The second tunic of the eye is composed of the choroid, ciliary body, and the iris. It lines the inner side of the sclerotic, and is perforated to allow the optic nerve to enter, and has a circular opening in front,

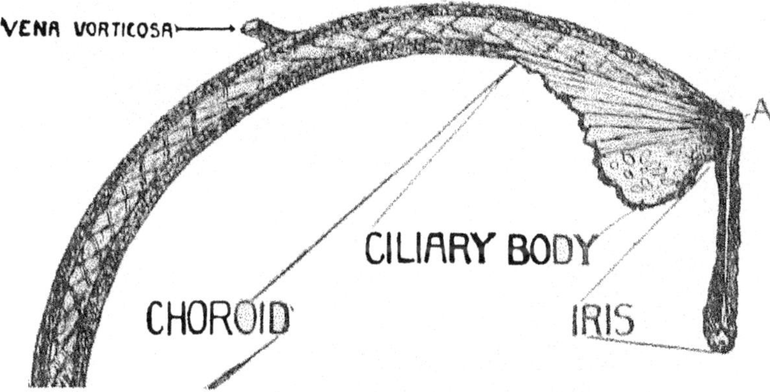

Cut showing Second Tunic.

which is known as the pupil. Through this tunic the eye obtains its principal blood and nerve supply. This is the tunic in which the pigment is deposited for the purpose of absorbing light. The choroid is said to nourish the retina and the vitreous. The ciliary muscles are within the ciliary body, and are used for accommodating. The iris is the most anterior portion of the second tunic. It is located in front of the crystalline lens, and separates the anterior and posterior chambers; it gives the eye its color, regulates the amount of light which enters, and prevents spherical aberration of the lens. The third tunic is the retina. It is a very delicate, transparent membrane, made up of ten layers, one of which is the layer of optic nerve fibers. These fibers pass through the lamina cribrosa at the optic disc, and flatten out more and more

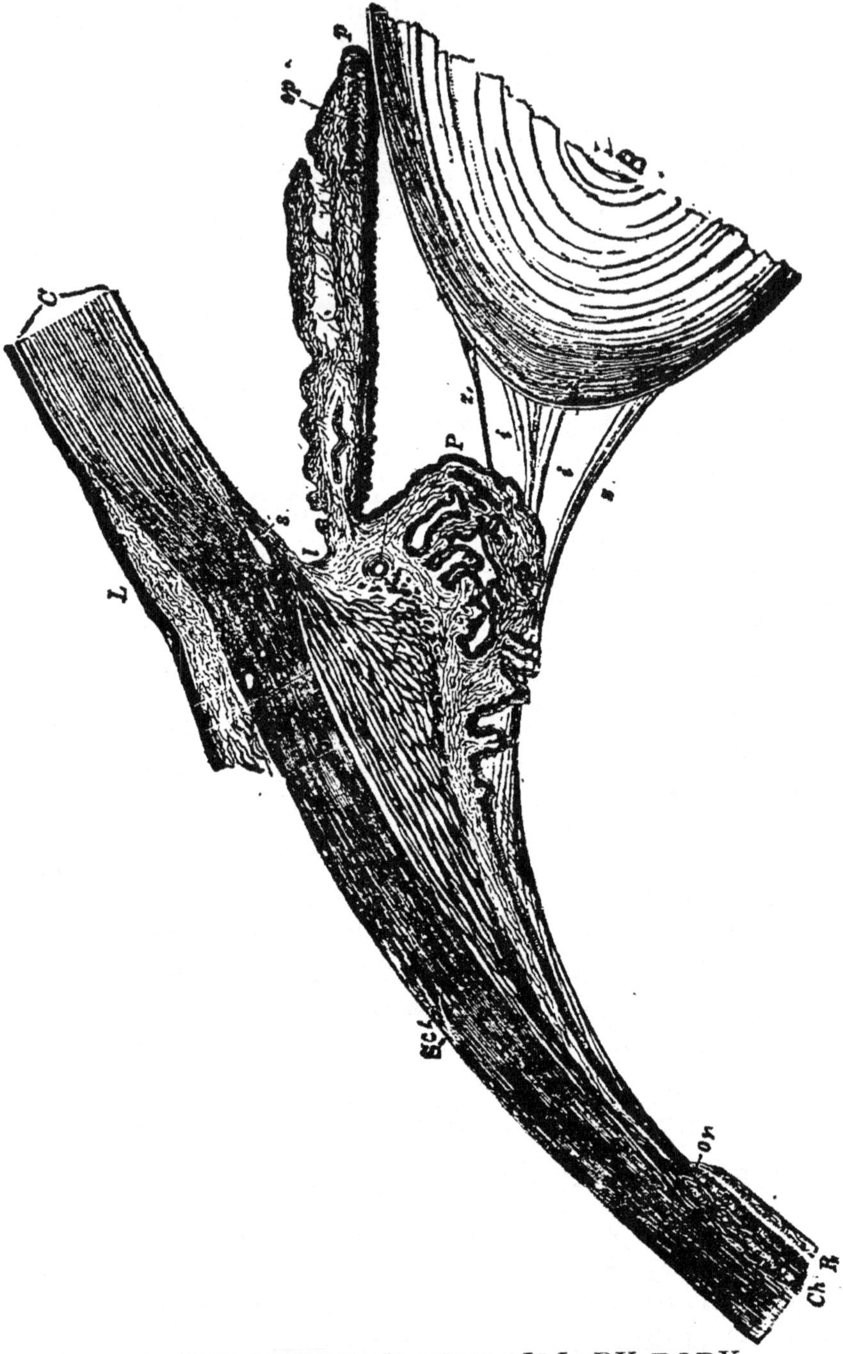

A CUT THROUGH THE CILIARY BODY.

C, Cornea; Scl, Sclerotic; Ch. Choroid; R, Retina; or, Ora Serrata; Z. Zonule of Zinn or Suspensory Ligaments; ii, Petit's Canal; p, Edge of Pupil; P, the most prominent part of the Ciliary Process; sp, Sphincter Pupillæ Muscle; s. Canal of Schlemm; L, Conjunctiva; B, Lens;

as they approach the front of the eye. The retina is attached in two places, at the optic disc and at its anterior border, the ora serrata. It is not attached to the choroid, but simply lies on it. In examining the retina with the ophthalmoscope you will notice the optic disc on the nasal side, which marks the entrance of the optic nerve into the globe. The macula lutea, which is the most sensitive spot of the retina (sometimes called the yellow spot, as it is said to turn yellow after death), is situated slightly on the temple side. The function of the retina is to receive the impressions of the waves of light and transmit them through the optic nerve to the brain. The space between the iris and cornea is known as the anterior chamber of the eye, and that between the iris and the lens, as the posterior chamber. Both of these chambers are filled with a transparent, watery fluid known as the aqueous humor. The large chamber back of the crystalline lens is known as the vitreous chamber, and contains the vitreous humor, which occupies a little more than three-fourths of the eyeball. It is a perfectly transparent substance, about the consistency of the white of an egg, and is enclosed in a thin transparent sac known as the hyaloid membrane. This membrane divides at the ciliary body and forms what is known as the anterior and posterior suspensory ligaments, which are attached to the lens capsule, thus forming what is known as Petit's Canal and the Zonule of Zinn. Within the lens capsule the crystalline lens is to be found. In shape the lens resembles a biconvexed lens, except that it is less curved in front than behind; in youth it is highly elastic, moderately firm, yet a perfectly transparent body, as clear as a crystal, and as we grow older it becomes harder and sometimes of a slightly straw tint. The crystalline lens is made up of layers closely resembling those of an onion, which accounts for its elasticity. The eyeball is imbedded in the fatty substance of the orbit, and is surrounded by a thin membranous sac, which isolates it and at the same time allows free movement. This sac is named the Capsule of Tenon. It is a very delicate membrane, consisting of two layers which invest the posterior part of the globe from the margin of the cornea backward to the entrance of the optic nerve, and is connected to it by a very delicate connective tissue. Both layers are lined on the inner surface by endothelial cells. The cavity between them is continuous with the space between the two layers of the sheath of the optic nerve, which is known as the subarachnoid space. The inner layer is known as the pia mater, and the outer as the dura mater, and between them empty the lymphatic vessels of the sclerotic. This capsule is penetrated by the (tendon) muscles of the eyeball near their insertion, which spread out fan shape and are attached to the sclerotic.

Anatomist (a-nat'-o-mist). A person who is skilled in anatomy.

Anatomy of Orbits. See Orbit.

Anatro'pia (Gr. ana = up + trope = a turning). That condition in which the eyes turn up.

Angle. A figure formed by two straight lines extending out from one point in different directions. **Acute Angle** is an angle less than a right angle, or less than 90 degrees. **Adjacent** or **Contiguous** angles are such as have one leg common to both angles. **Angle Alpha** is the angle formed by the optic and visual axis. **Angular Aperture** is the angle formed at the focal point of a microscope by the most divergent rays which enter the objective lens. **Angle of Convergence** is the angle which the two visual axes form in turning from their position of rest to a point less remote. The angle formed by the eyes turning from parallelism until the visual axes are directed to a point one meter distant on the median line, is called a meter angle of convergence and is the unit of the angle of convergence. When the visual axes meet on the median line, at a half meter distance, it is called a two-meter angle of convergence, and when looking at a third meter distance it is called a three-meter angle of convergence. **Critical Angle** is the angle beyond which a ray of light passing from a higher to a less refractive medium cannot emerge. **Angle of Deviation** is the angle formed between the direction of the incident and emergent rays, and shows the total deflection. **Angle of Gamma** is the angle formed at the center of rotation of the eye by the optic axis and a line drawn from the point on the object looked at. It varies with the refraction of the eye, and in emmetropia is about 5°; it increases in hyperopia and decreases in myopia. **Angle of Incidence** is the angle formed by the incident ray with the perpendicular at the surface. **Angle of Iris** is the angle between the iris and the cornea at the periphery of the anterior chamber of the eye. **Meter Angle** (for description, see under **Meter Angle**). **Minute Angle** is an angle formed by two straight lines that have traveled twenty feet distant from a point and separated the sixtieth part of a degree. **Oblique Angle** is an angle acute or obtuse, in opposition to a right angle. **Obtuse Angle** is an angle greater than a right angle or more than 90 degrees. **Optic Angle** is the angle included between the optic axes of the two eyes when directed to the same point. **Angle of Prism** is the angle formed by the surfaces of the prism which incline to each other. **Angle of Reflection** is an angle formed by the reflected ray with a line perpendicular to the reflecting surface, and is always equal to the Angle of Incidence. **Angle of Refraction** is the angle formed by the refracted ray with the perpendicular. **Right Angle** is an angle formed by a right line falling on another perpendicularly, or an angle of 90 degrees. **Angle of Strabismus** is the angle formed by the deviating eye with that of its fellow. **Visual Angle** is the angle formed between straight lines drawn from the boundaries of the object looked at and crossing the optical center of the eye.

Anian'thinopsy (Gr. an = not + ianthinos = violet colored + opsis = vision). Inability to distinguish violet shades.

Aniridia (an-ir-id'-e-ah). (Gr. an = not + iris.) Congenital absence of the iris.

Anisocoria (an-is-o-ko'-re-ah). (Gr. anisos = unequal + koro = pupil.) That condition where the two pupils are unequal.

Anisometropia (an-is-o-me-tro′-pe-ah). (Gr. an = not, iso = equal, metron = measure, ops = eye.) A difference of refraction in the two eyes. The defect is usually congenital, but it can be acquired, as in Aphakia, or operations of any kind. One eye may be emmetropic, the other hypermetropic, or myopic, or one more hypermetropic, myopic, or astigmatic than the other. When one eye is hypermetropic or emmetropic, and the other myopic, the hypermetropic or emmetropic eye is used for distance, and the myopic eye for near-vision.

Anisopia (an-is-o′-pe-ah). (Gr. an = not, iso = equal, ops = eye.) An inability of both eyes to receive equal impressions, not due to an unequal refractive state.

Anisosthenic (an-i-sos-then′-ik). (Gr. anisos = unequal + sthenos = strength.) Of an unequal strength, as when the muscles of the eyes are unequal in power.

Anisotropous, anisotropal, anisotropic (Gr. anison = unequal + tropos = a turning). Producing double refraction of a transmitted ray of light.

Ankyloblepharon (ang-kil-o-blef′-ar-on). Adhesions of the edges of the eyelids.

An′nular Muscle. A ring-shaped muscle (as the sphincter muscle of the iris).

Annulus (an′-nu-lus). A ring-shaped organ. A. ciliaris, boundary between iris and choroid.

Anomaly (a-nom′-a-ly). (Gr. anomalia = irregularity.) Deviation from normal standard.

Anoopsia (an-o-op′-se-ah). Where the eye has turned upward. (Strabismus.)

Anophthalmia (an-off-thal′-me-ah). (Gr. an = not + ophthalmos = eye.) Absence of the eyes.

Anopia (ano′pia). (Gr. an = without + ops = eye). Congenital absence of the eye, or eyes.

Anopsia (an-op′-se-ah). (Gr. an = not + opsis = sight.) Disuse of the eye from certain defects, such as cataract, amblyopia, and high refractive errors.

Anorthopia (an-or-tho′-pe-ah). (Gr. an = not + orthos = straight + ops = eye.) A defective state of vision which is unable to distinguish a want of parallelism.

Anterior (front part). Referring to the eye, the cornea would be the most anterior point.

Antimetropia (an-ti-me-tro′-pe-ah). (Gr. anti = opposite, metron = measure.) Where one eye is myopic and the other hypermetropic.

Antiseptic (an-ti-sep′-tik). (Gr. anti = against + septic = putrid.) A substance which is destructive to poisonous germs.

Aorta (a-or′-ta). (Gr. aeiro = to lift, raise.) The great main trunk of the arterial system, proceeding from the left ventricle of the heart, giving origin to all the arteries except the pulmonary (lungs).

Apex (a′-pex). (L., summit or tip.) The thin edge of a prism.

Aphacia. See Aphakia.

Aphakia (ah-fa′-ke-ah). (Gr. a = lacking, phakos = lens.) Absence of the crystalline lens.

Aphose (ah-foz′). (Gr. a = without + phos = light.) Any visual sensation due to absence or interruption of light.

Apical (a′-pik-al). Pertaining to the apex.

Aplanatic (ah-plan-at′-ik). (Gr. a = not + planetos = wandering.) That condition where there is neither spherical nor chromatic aberration, and the lines are also straight. (See Lens.)

Aponeurosis (ap-on-u-ro′-sis). (Gr. apo = from + neuron = sinew.) A fibrous band investing muscles and connecting them with tendons. It is white, fibrous, and destitute of nerves, and has very little blood.

Apparent Position. The position apparently occupied by an object seen through a refracting medium, as distinguished from its real position.

Appendages of the Eye are the orbits, the eyebrows, the eyelids, the conjunctiva, the lachrymal apparatus, the muscles, the aponeurosis, and vessels and nerves of the orbit.

Applanatio-corneae (ap-lan-a′-she-o-kor′-ne-e). A condition in which the cornea becomes flattened.

Applana′tion (L. ad = to + planare = to flatten). Flattening of a normally convex surface.

Approximation (a-prok-si-ma′-shon). (L. ad = to + proximare = to come near.) A continual approach to a true result. A result so near the truth as to be sufficient for a given purpose.

Aqueous Humor (a′-que-us hu′-mor). (L. aqua = water, humor = moist.) A transparent, watery fluid which fills the anterior and posterior chambers of the eye, the iris becoming the boundary line between the two chambers. Weight, 5 grains. Composed of water, albumen, and salts. The aqueous is a secretion formed by the epithelial cells lining the apices of the ciliary processes, passing through the pupil and leaving the eye through Schlemm's Canal to pass into the anterior ciliary vein. If this humor is allowed to escape it will re-form again. Its index of refraction is 1.33.

Aquocapsulitis (a'-kwo-caps-u-li'-tis). Serous inflammation of the iris.

Arachnoid (ar-ak'-noyd). (Gr. arachne = cobweb, eidos = resemblance.) Resembling a spider's web.

Arachnoid Membrane is a name given to several membranes which, by their extreme fineness, resemble spider webs. The delicate membrane between the dura mater and pia mater of the optic nerve and capsule of tenon.

Subarachnoid Space is located between the arachnoid membrane and the pia mater. It contains the greater part of the cerebro-spinal fluid.

Subdural Space is located between the dura mater and the arachnoid membrane, and contains a small amount of subdural fluid.

The use of the subarachnoid or cerebro-spinal fluid is to mechanically protect the nerve centers from shock, and to fill up space, as fat does in other parts of the body. Its amount is estimated at less than two ounces.

Arc (ark). (L. arcus = bow.) A portion of a curved line or segment of a circle.

Arcus senilis (ar'-kus sen'-il-es). (L. arcus = bow, senilis = old.) White circle in cornea near sclerotic. The arcus is first noticed as two small white opaque crescents, one above and the other below within the margin of the cornea, gradually extending until the ends join, forming a ring. The arcus will increase in opacity more than in width, and the cornea remains perfectly transparent within its circle and vision is not impaired by it. Cause, fatty degeneration. Condition in aged.

Area of Critical Definition. That portion of an optical image within which the detail is clearly defined.

Argamblyopia (ar-gam-ble-o′-pe-ah). Amblyopia from non-use of eye.

Argyll-Robertson Pupil. A pupil that will not respond to light, but contracts in accommodation. Can be seen in locomotor-ataxia.

Arithmetic (a-rith′-me-tik). (Gr. arithmos = number.) The theory of numbers, the art of computation, and the applications of numbers to science and business.

Artery (ar′-tery). (Gr. tube.) The vessel which carries the purified blood from the heart to the different cells of the body. The ophthalmic artery supplies the eye with blood and it arises from the internal carotid artery at the anterior clinoid process and is about one-twelfth of an inch in diameter. It enters the orbit through the optic foramen below and on the temple side of the optic nerve. It then passes over the nerve to the inner wall of the orbit and runs forward to the inner angle of the eye, where it divides into the frontal and nasal branches, which may be divided into an orbital group. It gives off the lachrymal, supra-orbital, anterior and posterior ethmoidal, palpebral, frontal and the nasal, muscular anterior, short and long ciliary, and the central artery of the retina.

The short ciliary arteries (six to ten in number) enter the eye through the sclerotic around the optic nerve to supply the choroid and ciliary processes.

The long ciliary arteries (two in number) pierce the sclerotic, one on each side of the eye, and run forward between the choroid and the sclerotic to the ciliary muscles, which they supply, and they join the anterior ciliary arteries.

The anterior ciliary arteries (five to eight in number) are derived from the muscular and lachrymal branches of the ophthalmic. They penetrate the sclerotic about a sixteenth of an inch from the corneal margin, and finally join the long posterior ciliary arteries to form the circulus arteriosus iridis major.

The arteria centralis retinae is the first and one of the smallest branches of the ophthalmic artery. It pierces the optic nerve about half an inch behind the eye ball and runs forward to the retina, in its centre, entering the eye through an opening in the lamina cribrosa, known as the porus opticus, where it divides into four branches.

Artificial Eye. A thin glass plate which resembles the sclerotic, cornea and iris. Artificial eyes are made in different sizes and colors, and are always fitted to match the sound eye. Before inserting the artificial eye it should be put into salt and water for a few minutes, then draw the upper lid out and down, and slip the eye up under; then draw the lower lid out and down, and in this way allow the eye to fall into posi-

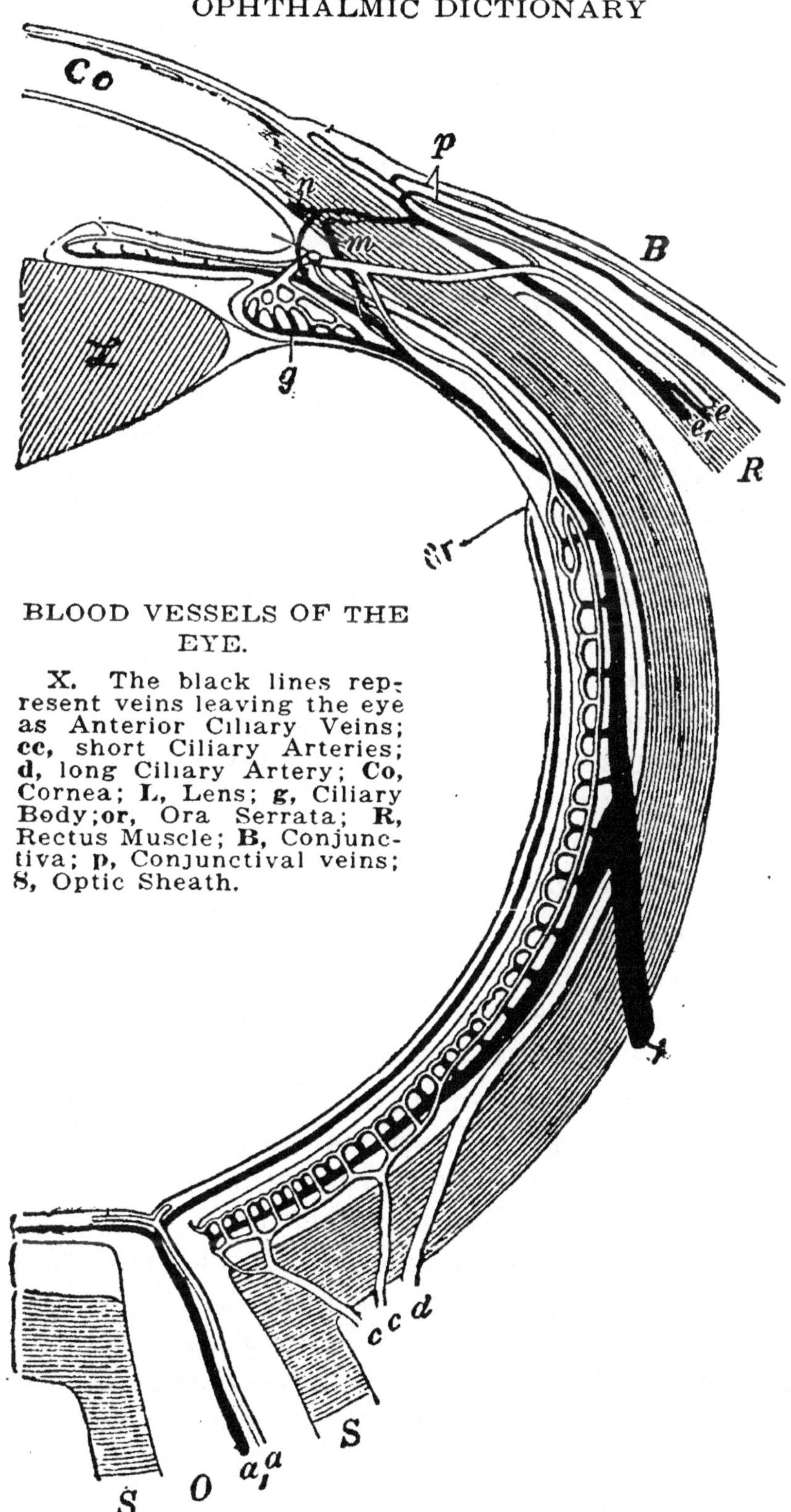

BLOOD VESSELS OF THE EYE.

X. The black lines represent veins leaving the eye as Anterior Ciliary Veins; **cc**, short Ciliary Arteries; **d**, long Ciliary Artery; **Co**, Cornea; **L**, Lens; **g**, Ciliary Body; **or**, Ora Serrata; **R**, Rectus Muscle; **B**, Conjunctiva; **p**, Conjunctival veins; **S**, Optic Sheath.

tion. As a rule, an artificial eye will last about a year, when it begins to lose its smoothness and a new one is required.

Artificial Pupil. One made by an operation (Iridectomy)

Asep'sis. Free from septic matter, or free from infection.

Asthenopia (as-then-o'-pi-ah). (Gr. a = lacking, sthenos = strength, ops = eye.) Weak, tired and painful vision; the result of straining some part of the mechanism of the eye. Vision may be normal. It is subdivided into three kinds: retinal, muscular, and accommodative.

Retinal. Where the eye cannot stand light without pain; intolerance of light; photophobia.

Muscular. A condition of the eyes in which the muscles controlling their movement suffer from speedy fatigue, causing pain.

Accommodative. Fatigue of the ciliary muscles by hypermetropia, presbyopia, or overwork in emmetropia.

Astigmagraph (as-tig'-ma-graf). An instrument used to demonstrate the state or condition of astigmatism of the eye.

Astigmatism (as-tig'-mat-ism). (Gr. a = not + stigma = a point.) Astigmatism is a term applied to an eye whose refraction is not the same in all its parts, causing the eye to focus the light at different points. It is subdivided into two kinds, **regular** and **irregular.** Irregular astigmatism is where there is a difference of refraction in one and the same meridian, and according to some authors, is subdivided into **normal** and **abnormal.** Normal irregular astigmatism is due in a great measure to irregularities in the refracting power of the different sectors of the lens, and causes a luminous point to appear stellate, or star shape. The abnormal variety is usually caused by ulcers, conical cornea, or injury of the cornea, but the same condition may be congenital. This kind of astigmatism cannot be corrected with lenses. Regular astigmatism is where we have the meridians of greatest and least curvature at right angles to each other, and are known as the principal meridians. This variety can be corrected with cylindrical lenses. It has five subdivisions, which merely serve to show the location of the foci, which are as follows:

No. 1. **Compound Hyperopic Astigmatism** is that condition in which the foci of the two principal meridians are back of the retina at different places when the eye is at rest and looking at infinity.

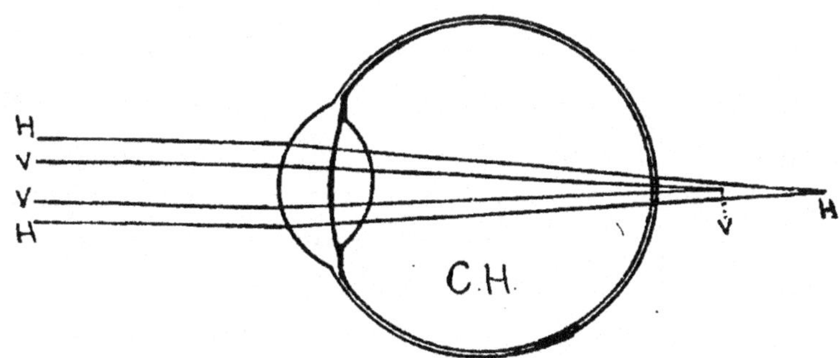

No. 1. Compound Hyperopic Astigmatism.

No. 2. **Simple Hyperopic Astigmatism** is that condition in which parallel rays enter the eye, and one of the principal meridians focuses on the retina, the other behind the retina, when the eye is at rest.

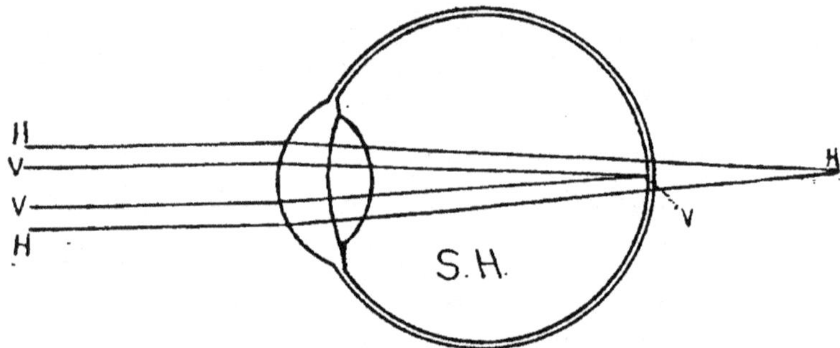

No. 2 Simple Hyperopic Astigmatism.

No. 3. **Compound Myopic Astigmatism** is that condition in which the two principal meridians focus in front of the retina at different places when the eye is at rest and looking at infinity.

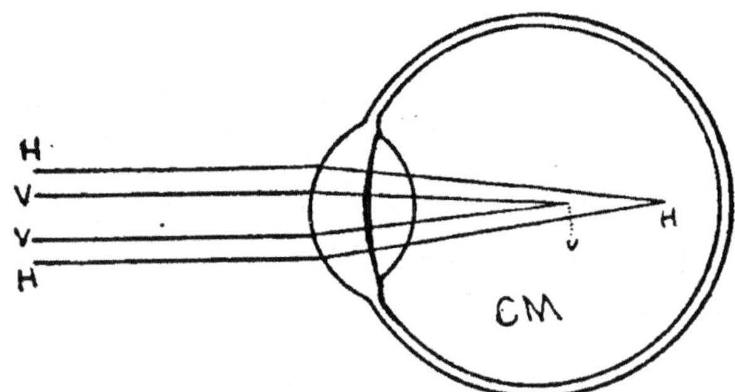

No. 3. Compound Myopic Astigmatism.

No. 4. **Simple Myopic Astigmatism** is that condition in which one of the principal meridians focuses on the retina and the other in front with the eye at rest and looking at infinity.

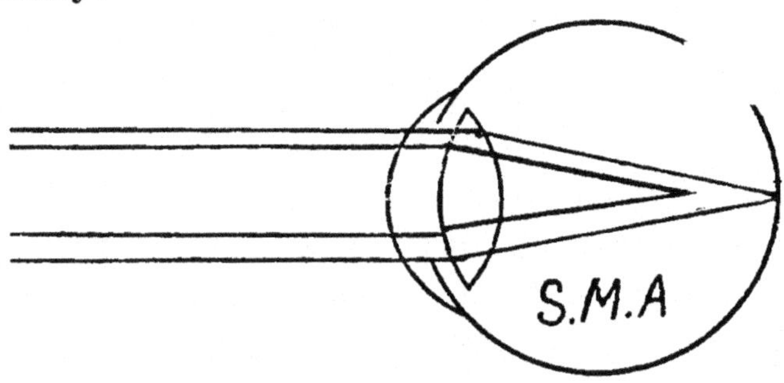

No. 4 Simple Myopic Astigmatism.

No. 5. **Mixed Astigmatism** is that condition in which one of the principal meridians focuses in front of the retina and the other behind the retina when the eye is at rest and looking at infinity. It derives its name, mixed astigmatism, from the fact that one meridian is hyperopic and the other myopic.

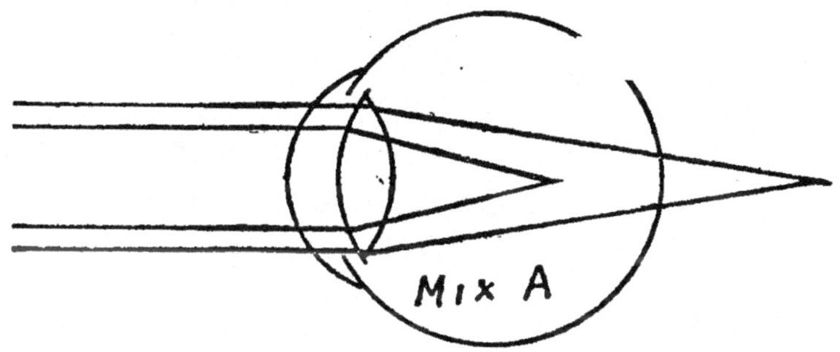

No. 5. Mixed Astigmatism.

Corneal Astigmatism is caused by irregularity of the curvature of the cornea.

Lenticular Astigmatism is caused by an irregularity of the curvature of the crystalline lens.

Astigmatism with the Rule is where the axis of a minus cylinder that will correct the astigmatism is nearer the 180th meridian than the 90th, or the axis of a plus cylinder that will correct the astigmatism is nearer 90 than 180; otherwise, it is against the rule.

Correcting cases of high astigmatism often proves unsatisfactory at the time. When the image is formed on the retina of such an eye it is much blurred at one of the principal meridians, and sometimes distorted. However, the patient accepts this, as his vision has never been better. So much so that when the correct lenses form a distinct retinal image, he fails to recognize it, and will sometimes say that the object looks distorted, as the fibers of his optic nerve are somewhat amblyopic, and therefore different from those of other people. In such cases the brain is in the habit of accepting vision from parts of the retina that are most distinct, and when wearing their correction for the first time, the vision shows very little improvement, if any. It is not uncommon for cases of say 4-D., of astigmatism to see very little better with their glasses at the time of fitting, but if they are worn persistently the vision is expected to improve in a few months.

Astigmom'eter. An instrument for measuring astigmatism.

Astig'mometry. The study of measurement of astigmatism.

Astringent (as-trin'-jent). (L. astringere = to contract.) An agent that causes contraction and arrests discharges.

Asymmetry (ah-sim'-et-re). (Gr. a = not + syn = with + metron = measure.) When the eyes do not correspond, or resemble each other in appearance, they are said to lack symmetry.

Atax'ia. (Gr. a = not + taxis = order.) Failure of muscles to coordinate.

Atom (at'-om). (Physics.) An ultimate indivisible particle of matter.

Atonic (aton'-ic). (Gr. atonia = languor.) Relaxed; without normal tone or tension.

Atrophy. (Gr. a = not + trope = nourishment.) A wasting away of a part from a lack of nutrition.

Atropine (at'-ro-pin). A poisonous alkaloid extracted from atropa-belladonna, or deadly nightshade; it temporarily paralyzes the sphincter iridis (circular muscles of the iris) and the ciliary muscles within the eye, consequently the pupil is dilated and accommodation fully relaxed, leaving the eye adjusted for its far point of vision in a state of rest. For this reason it is known as a **mydriatic** and a **cycloplegic**, and is used more than any other to suspend accommodation and dilate the pupil. Atropine paralyzes the sphincter muscle of the iris and the ciliary muscle, and hence results in dilatation of the pupil, and also an inability to see clearly near by. The dilatation of the pupil is a maximum one. If, in the case of a dilatation of the pupil caused by oculomotor paralysis, atrópine is instilled, the pupil becomes still more dilated. This proves that atropine, besides producing paralysis of the contracting fibers, causes also stimulation of the dilating fibers. The effect of the atropine makes its appearance in from ten to fifteen minutes after the instillation, and soon reaches its maximum. Commencing with the third day it begins to decrease again, but does not disappear completely until after the lapse of a week. The instillation of atropine, therefore, causes the patient a disturbance of rather long duration, and hence should be employed only when there is good reason for it.

Atropinism (at′-ro-pin-ism). A condition produced by the use of atropine.

At′ropinize. To put under the influence of atropine.

Atypic Hypermetropia (at-ip′-ik). Irregular Hypermetropia caused by tumors behind the eye, exerting such a pressure on the posterior pole that the region of the macula is pushed in front of the principal focus, the eye thus becoming hyperopic. It may be caused by detachment of the retina in the region of the macula.

Atypic Myopia (at-ip′-ik). Progressive Myopia, caused by the elongation of the eye.

Autophthalmoscope (au-tof-thal′-mo-skope). An ophthalmoscope planned in such a way that a person can examine his own eyes.

Average (av′-e-raj). Etymology obscure. The result of adding several quantities and dividing the sum by the number of quantities.

Avoirdupois (av′-or-du-poiz′). (Fr. aver = goods + de = of = pois = weight.) A system of weight in which 1 lb. = 16 oz. = approximately 7,000 troy grains.

Axial. Of, or pertaining to, an axis.

Axial Ametropia. Ametropia that is caused by the length of the eyeball on the optic axis.

Axially. In the direction of the axis.

Axiom (ak′-si-om). (Gr. axioma = a requisite, a self-evident principle.) A simple statement, of a general nature, so obvious that its truth may be taken for granted.

Axis (aks′-is). A straight line, real or imaginary, passing through the center of a body, on which it may revolve.

Optic A. An imaginary line through all centers of the eye from before back; that is, the center of the cornea, through the nodal points to the inner side of the macula lutea.

Visual A. A line from the macula lutea through the optical center of the eye to the point on the object looked at.

Major A. The longest diameter of the cornea or lens. **Minor A** the shortest.

Principal A. A line which passes through the optical center of a lens at right angles to both sides, it is the only ray of light to pass through unrefracted.

Secondary A. Any line crossing the principal axis at the optical center. These rays are refracted, but emerge in the same direction as they entered.

Axis of a Cylinder. The only meridian of a cylindrical lens without focusing power.

Axis of Mirror. A line which strikes the center of curvature at right angles to the surface is called its axis.

Axis of Refraction. The normal to the surface of a refracting medium at the point of incidence of a ray of light.

Axometer (ax-om′-e-ter). An instrument for the determination of the axis and focus of spherical, cylindrical or sphero-cylindrical lenses.

BACILLAR LAYER (bas′-il-ar). The layer of rods and cones of the retina.

Barom′-eter. (Gr. baros = weight + metron = measure.) An instrument indicating the atmospheric pressure.

Basalis Lamina, or membrane of Bruch. The membrane which separates the choroid from the pigmentary layer of the retina.

Base. (L. bassus = low.) (a) The line or surface forming that part of a figure on which it is supposed to stand. (b) The base of a system of logarithms is the number which, raised to the power indicated by the logarithm, gives the number to which the logarithm belongs. (c) In percentage, the number which is multiplied by the rate to produce the percentage. The base of a prism is its thickest part.

Base Apex Line. A straight line drawn from the thickest to the thinnest part of a prism.

Base Curve. The meridian of least refraction, on the toric side of a lens.

Basedow's Disease. (See exophthalmic goiter.)

Beer's Knife. A knife with a triangular blade for corneal incision.

Bi. (L. bis = twice.) Is employed to signify two things in one; for instance, bifocal, biconcave, biconvex.

Biconcave. Concave on both sides.

Biconvex. Convex on both sides.

Bifocal (bi-fo'-kal). Having a double focus. A lens with two focal lengths, having two parts, the upper for distance and the lower for near vision. They can be made up in five different ways, namely:

- No. 1. **One-piece Bifocal.**
- No. 2. **Split or Franklin Bifocal.**
- No. 3. **Perfection Bifocal.**
- No. 4. **Cemented Bifocal.**
- No. 5. **Invisible Bifocal.**

The one-piece bifocal is made of glass with the same density throughout. Its double focus is governed by change of curvature. It can be mounted with or without rims. The split or Franklin and perfection bifocals must be worn in rims, while the cemented and also invisible bifocals can be worn without rims. The invisible bifocal lenses are made by fusing under intense heat, two pieces of curved glass with different densities. A bifocal lens consists of two parts of two different foci. In hypermetropia with presbyopia, or old sight, the upper is the weaker for distance, the lower being stronger for near objects. In myopia, the upper should be the stronger and the lower the weaker glass. In this way the patient has good distant vision without the extra strain on the accommodation.

Canada Balsam is used in cementing bifocal lenses in the following manner: First be sure that the lenses are **perfectly clean.** Then squeeze a small drop of the balsam on to the large lens, and press the scale upon the balsam until it spreads out thoroughly between the glasses, being careful not to break the lenses. Then place the lenses on a piece of metal over a small flame, and heat them slowly until all the bubbles disappear, and until the balsam is nearly hard—just about hard enough to take a slight impression of the finger nail. It is impossible for a novice to accurately judge just how much to heat the lenses, but with practice it becomes a simple matter. The success depends largely on their being heated just long enough. If they are not heated enough they will slide out of position, and if they are heated too much they will chip off very easily.

Binocular. (L. bis = twice, oculus = eye.) Pertaining to both eyes. In vision it refers to the ability of both eyes to see the same point of an object at the same time. To prove it exists a 5° prism placed base up or down over one eye should make the object looked at appear double.

Biorbital Angle. The same as the optic angle.

Birefractive. Doubly refractive.

Bi-Spherical. A lens with a sphere on both sides.

Blear-eye. Marginal blepharitis.

Blennorrhea (blen-or-e′-ah). (Gr. blennos = mucus + rhoia = a flow.) Excessive mucous discharge.

Blepharadenitis (blef-ar-ad-en-i′-tis). (Gr. blepharon = eyelid + aden = gland + itis.) Inflammation of the meibomian glands.

Blepharal (blef′-ar-al). (Gr. blepharon = eyelid.) Pertaining to the eyelids.

Blepharelosis (blef-ar-el-o′-sis). Ingrowing eyelashes. (See Trichiasis.)

Blepharism (blef′-ar-ism). Where there is an inability on the part of the patient to refrain from winking. (Blinking.)

Blepharitis (blef-ar-i′-tis). (Gr. blephron = eyelid + itis = inflammation.) Inflammation of the eyelids. **Ciliaris or marginalis b.** That condition where the hair follicles of the eyelid are inflamed. **Squamosa b.** A marginal blepharitis in which the edges of the lids become scaly. **Uncerosa b.** An ulcerous form of marginal blepharitis.

Blepharochalasis (blef′-ar-o-kal′-as-is). (Gr. blepharon = eyelid + chalasis = a slackening.) Relaxation of the skin of the eyelid, due to atrophy of the intercellular tissue. A condition in which folds of the skin hang down.

Blepharoclonus (blef-ar-ok-lo-nus). (Gr. blepharon = eyelid + klonos = a tumult.) Clonic spasm of the eyelids.

Blepharoconjunctivitis. Inflammation of the eyelids and conjunctiva.

Blepharoncus (blef-ar-ong′-kus). (Gr. blepharon = eyelid + onkos = a tumor.) A tumor or swelling of the eyelid.

Blepharoplegia (blef-ar-o-ple′-ge-ia). (Gr. blepharon = eyelid + plege = stroke.) That state in which the eyelid is paralyzed, causing ptosis.

Blepharoptosis (blef-ar-op-to′-sis). (Gr. blepharon = eyelid + ptosis = a falling.) That condition where the upper eyelid droops from paralysis.

Blepharopyorrhea. (Gr. blepharon = eyelid + pyon = pus + rhoia = I flow.) A discharge of mucus from the eyelids.

Blepharorrhaphy (blef-ar-or′-af-e). (Gr. blepharon = eyelid + rhaphe = seam.) The stitching together of a part of the edges of a slit between the eyelids.

Blepharospasm (blef′-ar-o-spasm). A spasmodic contraction of the orbicularis palpebrarum muscle, so that the lids are firmly pressed against the globe. Occurs where photophobia is marked. It is reflex from the irritation of the fifth nerve, and occurs in neuralgia of its branches; in inflammation of the conjunctiva or cornea; from foreign bodies; errors of refraction, etc.

Blepharostat (blef-ar′-o-stat). (Gr. blepharon = eyelid + statos = fixed.) An instrument used for holding the eyelids apart.

Blepharostenosis (blef-ar-o-ste-no′-sis). Gr. blepharon = eyelid + stenosis = a narrowing.) A narrowing of the palpebral slit between the eyelids.

Blepharosynechia (blef-ar-o-sin-ek′-i-a). (Gr. blepharon = eyelid + synechia = continuity.) A condition in which there is a growing together of the eyelids.

Blepharotomy (blef-ar-ot′-o-me). (Gr. blepharon = eyelid + tome = incision.) A surgical operation for the cutting of the eyelid.

Blind. Loss of sight. Day-blindness is where vision is better at night. Night-blindness is defective vision at night-time.

Blind Spot. Also known as the optic disc, or optic papilla. It marks the entrance of the optic nerve on the retina. Not sensitive to light.

Blinking. That condition in which there is an involuntary winking.

Blood (blud). A red, slightly translucent fluid which circulates in the principal vascular system of animals, carrying nourishment to all parts of the body, and bringing away waste products to be excreted. In the veins its color is somewhat darker, owing to the loss of oxygen while passing through the tissues.

Blood Vessels. (See veins and arteries.)

Bonnet's Capsule. The same as Tenon's Capsule.

Bowman's Membrane. The second anterior layer of the cornea.

Brachymetropia (brach-e-me-tro′-pe-a). (Gr. brachus = short, metron = measure, ops = eye.) The same as myopia and hypometropia. It is an eye where parallel rays of light will focus in front of the retina with the muscles of accommodation at rest.

Brain. The organ of intellect. A nervous mass within the skull divided into many parts.

Brow-Ague. Supra-orbital neuralgia; a superficial pain in the region of distribution of the first division of the fifth nerve.

Bruch's Glands. The lymph-follicles of the conjuctiva of the lower eyelid.

Bruch's Membrane. The inner layer of the choroid coat of the eye.

Bruch's Muscle. (See Ciliary Muscle.)

Buphthalmia (buf-thal′-me-ah). (Gr. bous = ox + ophthalmos = eye.) Enlargement of the eye.

Buphthalmus. (See Buphthalmia.)

CAMPIMETER (kam-pim′-e-ter). (L. campus = field + Gr. metron = measure.) An instrument for measuring the field of vision, replaced by perimeter.

Canada Balsam. A liquid resin obtained from the balsam-fir tree, which grows in Canada. It is used for the purpose of cementing lenses together. It is easily melted if heated, and readily soluble in alcohol. (See bifocal for its use.) Index of refraction 1.52.

Canalic′ulus. A small canal or channel.

Canal of Cloquet. The name given to Stilling's Canal after Jules Germain Cloquet, Parisian surgeon, 1790-1883.

Canal of Petit. The space which surrounds the crystalline lens between the suspensory ligaments.

Canal of Schlemm. Circular canal surrounding the eye at sclerocorneal junction. (See Anatomy.)

Canal of Stilling. The canal which runs through the vitreous humor from the entrance of the optic nerve to the posterior surface of the lens. It is lined by the hyaloid membrane. This canal is said to convey the minute artery from the central artery of the retina to the back of the lens, during development of the eye. The artery then disappears, but the canal remains. It is also known as the **Hyaloid** Canal.

Canals of Fontana. A number of little spaces or openings between the iris and cornea, in the sclerotic. (See Spaces of Fontana; also see Schlemm's Canal.)

Cancel (kan′-sel). (L. cancelli = a lattice.) Originally to draw lines across a calculation. To strike out or eliminate as a common factor in the terms of a fraction, a common term in the two members of an equation, etc.

Canthectomy (kan-thek′-to-my). (Gr. kanthos, canthus + ektome = excision.) An operation in which part of the canthus is cut away.

Canthitis (kan-thi′-tis). Inflammation of the angles of the eyelids.

Cantholysis. (Gr. kanthos, canthus + lysis =

loosening.) Incision of the canthus to widen the slit between the lids.

Canthoplasty (kan′-tho-plas-te). A surgical operation for lessening the pressure and friction of the upper lid by cutting the outer canthus. **Plastic c.** operation, an operation for restoring a lost part.

Canthotomy (kan-thot-′o-me). An operation for the slitting of either canthus.

Can′thus. (Gr. "angle of the eye.") The angle at the junction of the eyelids, known as the inner and outer canthi.

Cap′illary (hair like). Any one of the little vessels which conduct the blood from the arteries to the veins.

Capsule (kap′-sule). (L. capsula = a box.) A sac which encloses an organ for the purpose of support, protection and lubrication. The capsule of the eye lens.

Capsule of Tenon. (See Tenon's Capsule.)

Capsulitis (kap-su-li′-tis). Inflammation of the capsule of the crystalline lens.

Capsulotomy (kap-su-lot′-o-my). An operation for the cutting of a capsule, as that of the lens.

Cardinal Points (that on which a thing turns or depends). Points which play an important part in the course of light through a spherical surface. There are four in number. The two principal foci and the two nodal points. The first principal focus is the point from which light rays emanate and pass through a spherical lens and emerge parallel to its principal axis. The second principal focus is the point where the emergent rays cross each other when the incident rays have been parallel to the principal axis. (See Nodal Point.)

Carotid (ka-rot'-id). (Gr. karos = deep sleep, stupor; so named from the effect of pressing on them.) The carotid arteries or carotids are the two great arteries of the neck, that convey the blood from the aorta to the head, brain and eye. The common carotids, one on either side of the neck, divide into an external and an internal branch, the former supplying the exterior of the skull.

Cartilage (kar'-til-aj). (L. Cartilago = gristle.) The gristle or white elastic substance in different parts of the body. (See Tarsus.)

Cartilaginous (kar-til-aj'-in-us). Relating to or consisting of cartilage.

Caruncula Lachrymalis (kar-un'-ku-lah). (L. a small fleshy mass. L. lacrima = a tear.) Is the small reddish body at the inner canthus of the eye.

Cast. A cast in the eye would apply to strabismus.

Cataphoria (kat-a-fo'-re-ah). (Gr. kata = down, phoria = "tending.") That condition in which one of the eyes, though parallel with its fellow when in use, turns downward when the extrinsic muscles are in a state of rest. **Esocataphoria** is the tendency of the visual line inward and downward. **Exocataphoria** is a tendency of the visual line outward and downward.

Cataract (kat'-ar-akt). (Gr. kataraktes = to break down.) Any opacity of the crystalline lens or lens capsule of the eye. **Lenticular c.**, an opacity of the lens proper. **Capsular c.**, an opacity of the lens capsule. **Senile c.**, an opacity of the lens due to age. **Traumatic c.**, a cataract due to an injury. **Pyramidal c.**, an opacity in the center, yet at the anterior pole, of the lens. **Secondary c.**, a cataract appearing after the extraction of the lens, caused by that part of the lens capsule still attached to the hyaloid membrane becoming opaque. **Cortical c.**, that condition in which the border or outer layers of the lens are losing their transparency. **Hard c.** (See Senile c.). **Soft c.**, where the lens is soft and milky. **Polar c.**, an opacity confined to the anterior or posterior pole of the lens.

Catacaustic. (See Caustic surface.)

Cat'adioptrical. (Physics.) Pertaining to, produced by, or involving, both the reflection and refraction of light.

Catopter (kat-op'-ter). (Gr. kata = down + optomai = I see.) A reflecting optical glass or instrument; a mirror.

Catoptric (kat-op'-trik). Relating to that branch of optics called catoptrics; pertaining to incident and reflected light. The whole doctrine of catoptrics is founded on this simple principle that the angle of reflection is equal to the angle of incidence.

Catoptric Test (kat-op'-trik). A test for cataract by light reflected from the crystalline lens. In this test ask the patient to look straight ahead, then hold a lighted candle about twelve inches in front of the eye, a little to one side, while you stand slightly on the other and look into his pupil. If there is no opacity of the lens or capsule you will notice three images of the candle. The first will be on the surface of the cornea in an upright position, the second will be on the anterior surface of the lens, also upright, while the third will be inverted and much smaller on the posterior surface of the lens, but when there is a cataract you will fail to find the inverted image (known as Pyrkinges Images).

Catoptry (kat-op′-tre). The unit of reflective power of curved mirrors. A mirror that will reflect parallel rays of light to a point of focus at a distance of one meter.

Cat's-eye Pupil. Where the pupil of the eye is long and narrow (slit-like).

Caustic Curve (kaus′-tik). A curve to which the rays of light reflected or refracted by another curve are tangent.

Caustic Surface. A surface to which rays of light reflected or refracted by another surface are tangents. Caustic curves are called catacaustic when formed by reflection and diacaustic when formed by refraction.

Cellulitis (sel-u-li′-tis). Inflammation of the loose tissues of the orbit.

Center (of curvature). If the surface of a lens were completed so as to form a circle, its center would be the center of curvature. (See Optical Center.)

Centimeter (sen′-tim-e-ter). One-hundredth part of a meter.

Centrad (sen′-trad). Toward the center; unit of measurement for prisms which will produce a deviation in a ray of light one-hundredth of a radian.

Centric (sen′-trik). Pertaining to a nerve center.

Centrifugal (sen-trif′-u-gal). Tending, or causing, to recede from the center.

Centrifugal Impression. An impression sent from a nerve center outwards to a muscle or muscles by which motion is produced.

Centrophose. (Gr. kentron = center + phos = light.) A subjective sensation of a light spot or patch, the cause being located in the optic brain center.

Ceratitis (ser-at-i′-tis). (Gr. "horn" + suffix, itis = inflammation.) The same as keratitis.

Ceratonosus (ser-at-on′-o-sus). (Gr. keras = horn + nosos = disease.) Any disease of the cornea.

Ceratotome (se-rat′-o-tom). A knife for dividing the cornea.

Chalazion (chal-a′-zi-on). (Gr. "Hail.") A tumor on the eyelid. On the under surface of the tarsal plate of the upper and lower lid are numerous creases or depressions running at right angles to the margin of the lid. There are about thirty of them in the upper lid and about twenty in the lower. In these depressions are small tubular glands, called meibomian glands, and their ducts open next to the margin of the lid. A chalazion is an enlargement of one or more of these glands, due to the stoppage of their ducts, and is usually chronic in character. A chalazion is also called a tarsal tumor, tarsal cyst, or meibomian cyst, etc. It is not a true retention cyst, but its contents may soften so that it will become an encysted abscess. At first its contents are gelatinous, but later may become purulent. The tumor is firm, round, with the skin moving freely over the mass, but it is firmly attached to the tarsal plate. It has so much the appearance of a sebaceous cyst that one is liable to be mistaken in the diagnosis, unless he is familiar with the disease. Usually chalazion tends toward the conjunctiva, and, if the lid is everted, the position of the tumor may be located by a bluish discoloration, or, if the contents are purulent, a yellowish discoloration. The primary cause of this trouble is not definitely known, but a debilitated condition of the system, eye-strain, and blepharitis marginalis seem to be the factors in producing chalazion.

Chamber. (Gr. "vaulted room.") The spaces of the eye. A term used in speaking of the eye, in which there are three chambers. The **anterior c.** is the space between the cornea and the front surface of the iris; the **posterior c.**, the space between the iris and the front surface of the lens capsule. These two are filled with the aqueous humor and communicates through the pupil; **vitreous c.**, the space surrounded by the hyaloid membrane behind the lens, occupying four-fifths of the globe of the eye.

Chemosis (ke-mo′-sis). (Gr. "an aperture" and suffix osis, signifying "morbid condition.") A swollen condition of the conjunctiva, forming an elevated ring around the cornea.

Chiasm (ki′-asm). (Gr. chiasma = two crossing lines.) A crossing; especially the crossing of the fibers of the optic nerve (optic commissure).

Chiastometer (ki-as-tom′-e-ter). An instrument for ascertaining the deviation of the optic axis.

Chlorophane (klo′-ro-fan). (Gr. chloros = greenish yellow + phaino = I show). A green-yellow pigment from the retina.

Chloropsia. (Gr. chloros = yellowish green + opsis = eyesight). Green vision. A condition in which all objects appear to be colored green.

Choked Disc. Congested and inflamed state of the optic disc.

Chondral (kon′-dral). (Gr. chondros = cartilage.) Pertaining to cartilage.

Chorea (ko-re′-ah). (Gr. choreia = choral dance.) St. Vitus' dance. A disease of the nervous system, characterized by a succession of irregular, clonic involuntary movements, of limited range occurring in almost all parts of the body. Control of the muscles is not lost, but voluntary motions are interfered with by the involuntary contractions.

Chorioid. See Choroid.

Choroid (ko′-roid). "Skin-like bag with multitude of blood vessels." That part of the second tunic of the eye extending from the optic nerve entrance to the posterior limit of the ciliary body. It is covered by the retina on the inner side.

It constitutes the posterior two-thirds of this vascular tunic. Its thickness gradually diminishes toward the ciliary body. The choroid consists of a compact mass of connective tissue, stroma, which supports the numerous blood vessels of varying size; forming three layers.

1. The layer of stroma containing large blood vessels.

2. The chorio-capillaris.

3. The membrana vitrea.

The first layer is the most conspicuous of the three, for its large blood vessels emerge into four main trunks, the vena vorticosae; these pierce the sclerotic at equal distances apart, running obliquely backward. The nerves of the choroid are derived from branches of the long and short ciliary nerves. Its blood supply is from the short posterior, long posterior and anterior ciliary arteries. The red reflex seen in the eye when viewed with the retinoscope and ophthalmoscope is due to the reddish color of the blood vessels in the choroid showing through the retina.

Choroidal Fissure. The opening in the choroid through which the optic nerve passes to form the retina.

Choroideremia (ko-roi-de-re′-me-ah). Absence of the choroid.

Choroiditis (ko-roi-di′-tis). Inflammation of the choroid.

Choroidocycli′tis. Inflammation of the choroid and ciliary processes.

Choroidoiritis (ko-roi-do-i-ri′-tis). Inflammation of the choroid and iris.

Choroidoretini′tis. Inflammation of the choroid and retina.

Chromatic (kro-mat′-ik). (Gr. chroma = color.) Relating to color.

Chromatic Aberration. See Aberration.

Chromatodysopia (kro-mat-o-dys-o′-pi-ah). (Gr. chroma = color + dys = bad + ops = eye.) Color-blindness.

Chromatogenous (kro-mat-oj′-en-us). (Gr. chroma = color + gennao = I produce.) Producing color.

Chromatology (kro-mat-ol′-o-gy). The study of colors.

Chromatom′eter. An instrument for measuring color or color perception.

Chromatophobia (kro-mat-o-fo′-be-ah). (Gr. chroma = color + phobos = fear.) An abnormal fear of color.

Chromatopsia (kro-mat-op′-se-ah). (Gr. chroma = color + opsis = vision.) Abnormal sensation of color, due to disorders of the optic centers, or to drugs, especially santonin.

Chromatoptometry (kro-mat-op-tom′-et-ry). Taking the measurement of the power of color perception.

Chromometer (kro-mom′-et-er). An instrument for measuring coloring matter present.

Chromoptometer (kro-mop-tom′-et-er). An instrument to test the color sense.

Chromoscope. (Gr. chromo = color + skopeo = I view.) An apparatus for testing the color sense.

Cibisitome (sib-is′-it-om). An instrument for incising the lens capsule.

Cilia. (L. cilium = eyelash.) The eyelashes. Hair.

Ciliariscope (sil-i-ar′-is-cope). An instrument for examining the ciliary region of the eye.

Ciliary (sil′-i-a-ry). Pertaining to, or like, the eyelashes.

Ciliary Body. The middle part of the second tunic, composed of ciliary processes, ciliary veins, ciliary muscles, ciliary nerves and arteries.

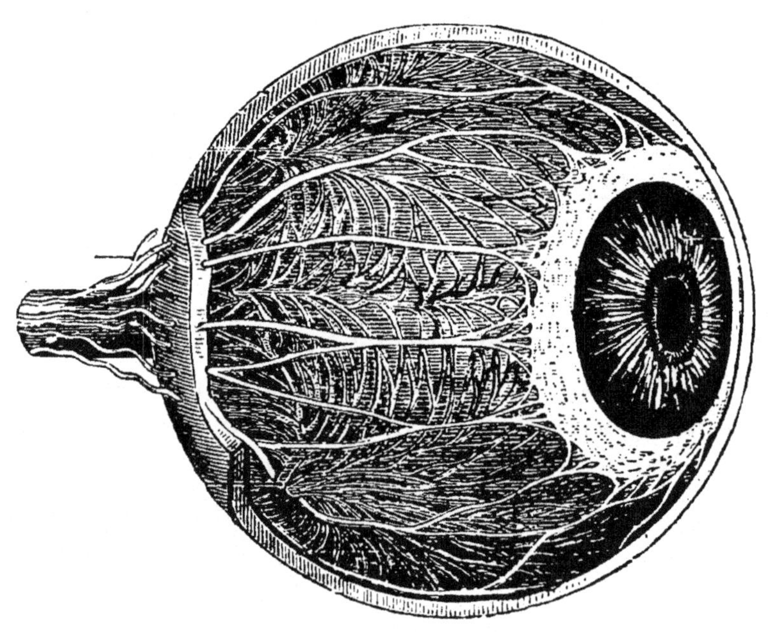

Cut showing Choroid. Ciliary Body. Iris and Nerves

Ciliary Processes. The radiating circular folds composed of a connective tissue stroma, which pass up over the ciliary body. There are about sixty or seventy in number.

Cillo (cil′-lo) or **Cillosis.** (Gr. "I move.") A trembling or spasmodic twitch of the eyelids.

Cinerea (sin-e′-re-ah). (L. cinereus = ashy.) The gray matter of the nervous system.

Circle (sir′-kl). (L. circulus, dim of circus. Gr. kirkos—a ring.) A plane figure whose periphery is everywhere equally distant from a point within it, the center

Circles of Haller. Venous and arterial circles of the eye.

Circulation (çir-cu-la'-shun). (L. circulare = to encompass.) The passage of blood in going from and returning to the heart after having made a circuit of the body.

Circum. (L. around.) A prefix denoting a circular movement, or a position surrounding the part indicated by the word to which it is joined. **Circumlental**, surrounding the crystalline lens; **circum ocular**, around the eye; **circumorbital**, around the orbit.

Circumference (ser-kum'-fe-rens). (L. circum = around + ferre = to bear.) The line which bounds a circle.

Clonic Spasm. (Gr. commotion.) An intermittent involuntary contraction of a muscle, which shows itself when the muscle is in use.

Cobalt. A hard, brittle, and heavy metal whose compounds afford pigments. The Cobalt-blue test glass being named after the blue pigment.

Cobalt-Blue Glass contains a great deal of red, and allows only the blue and red rays of the spectrum to pass through, neutralizing the other five colors contained in white light. It is used in testing the ametropia when the patient is not color blind.

Cocaine (ko-ka'-in). A local anæsthetic and mydriatic. Cocaine dilates the pupil, and hence would seem to call for mention in this place, although, strictly speaking, it does not belong to the mydriatics proper—that is, the dilatation of the pupil by cocaine is not produced, as in their case, by its action upon the contracting or the dilating fibers of the iris, but by a contraction of the blood vessels of the iris. The dilatation of the pupil is therefore only a moderate one, and the reaction of the pupil to light persists; moreover, mydriatics and miotics still produce an effect. If cocaine is instilled into an eye the pupil of which has been dilated by atropine, the dilatation increases somewhat in consequence of the anaemia of the iris which then ensues; hence the mydriasis produced by the simultaneous action of atropine and cocaine is the most complete that can possibly be attained. The accommodation is not paralyzed by cocaine, but only somewhat weakened.

Cohesion (ko-he′-shun). (L. con = together + haereo = to stick.) That form of attraction bv which the particles of a body are united throughout the mass, whether like or unlike.

Colmascope (col′-ma-scope). An instrument for the detection of strains and stresses in lenses, either mounted or unmounted, as a result of undue tightening of screws or imperfections in manufacture.

Collyrium (col-lyr′-i-um). (Gr. kollyrion = an eyewater.) Any lotion to be dropped in the eye.

Coloboma (kol-o-bo′-mah). (Gr. koloboma = an imperfection.) A tear or break in the eyeball, as in the iris or choroid.

Color-Blindness (Achromatopsia). Blindness for one or more colors. Due to the absence from the retina of one or two of the three primary substances (according to Hering). The test is made by presenting the patient with samples of different colored yarns—a number of each color, but different shades—and the patient is requested to separate them. Persons having this anomaly of vision are generally unaware of it themselves.

Commissure (kom′-mis-ur). (Optic.) (L. committere, "to unite.") The x-like crossing of the optic nerves.

Composite Number (kom-poz′-it). (L. com = together + ponere = to put.) A number which can be exactly divided by a number exceeding unity.

Compound Lens. A lens that contains a sphere and a cylinder.

Comus (ko′-mus). A cone. A crescentic patch of atrophic choroid tissue near the optic papilla in myopia.

Concave. (L. concavus = hollow.) Hollow and curved or rounded. The opposite to convex.

Concavo-convex. Concave on one side and convex on the other. If the convexity exceeds the concavity it is known as a periscopic convex lens. If the concavity exceeds the convexity it is known as a periscopic concave lens.

Concentric (kon-sen′-tric). (L. con = together + centrum = center.) That which has a common center with something else.

Concentrical (tri-kal). Having a common center, as circles of different size, one within another.

Concomitant (kon-com′-it-ant). (L. concomitare = to accompany.) Accompanying. **Concomitant Squint** is a condition where the two eyes deviate, but accompany one another in their movement. The object can be seen by either eye, but not the two eyes at the same time.

Cone (kon). (Gr. konos = peg.) The elementary form considered in arithmetic is a solid generated by the revolution of a right-angled triangle about one of its sides as an axis.

Cone Muscle Test. This consists of a cone cemented to a ground glass disc, and is used as follows: It is inserted into one cell of trial frame in front of the correction for the ametropia, which must be properly centered as to pupillary distance, and a solid blank disc is put into the cell in front of the other eye. The patient's attention is then directed to a light (preferably a candle or small gas light) twenty feet away; and the action of the cone is such that the light will resolve itself into a circle of light. The other eye is then uncovered, and if there is no muscular error the light will appear in the center of the circle. If there is muscular error the light will be either above or below or to one side of center, and can be brought to center by the proper prism with base in proper position. This does away with necessity for computations where there is combined prismatic error in different angles, and gives at once the position of the base of the correcting prism. (See Muscular Imbalance.)

Conical (con′-ic-al). That which is round and tapering to a point.

Conical Cornea. A cone-like protrusion of the cornea anteriorly due to increased intraocular pressure and weakening of its central portion.

Many forms of operations have been suggested and while some have been of slight benefit, no complete cure is known. Vision can usually be improved by means of an opaque disc with an opening in its center and worn as an eye glass. (Same as Keratoconus.)

Conjugate. (L. con = with + jugare = to join.) Coupled. To yoke together.

Conjugate Deviation. The deviation of both eyes in the same direction.

Conjugate Foci (kon′-ju-gat). Two points so situated in relation to each other that the direction of a ray proceeding from either of them, after reflection or refraction, passes through the other. **Secondary c.** A conjugate foci formed on a secondary axis.

Conjunctiva (kon-junc-ti′-va). (L. con = with + jungere = to join together.) The mucous membrane of the eye. It covers the anterior surface of the eye, passing backward about one-half inch, where it turns (fornix) to line the inner surface of the eyelids, thus forming a complete sac when the eyelids are closed. It consists of two portions; the ocular portion, which covers the sclerotic, and the palpebral portion, which lines the inner surface of the lids. The ocular portion is loosely connected with the sclerotic, but firmly attached to the edge (limbus) of the cornea. The palpebral portion is thick, opaque, highly vascular, and covered with numerous papillae. At the margin of the lids, it becomes continuous with the lining membrane of the ducts of the meibomian glands. At its outer and upper angle there are from seven to ten ducts through which the tears pass from the lacrimal glands. It receives its blood supply from the palpebral and lacrimal arteries. Its vein, the post-tarsal, is tributary to the ophthalmic vein. It receives its nerve supply from the fifth pair entering at the inner and outer angle of the orbit.

Conjunctivitis (kon-junc-tiv-i′-tis). Inflammation of the conjunctiva.

Convergence (kon-ver′-gence). (L. con = with + vergere = to turn.) The act or power of turning the eyes from their position of rest. When the eyes are emmetropic and orthophoric, the two functions, accommodation and convergence, work together, yet their objects are totally different, but their harmonious cooperation is none the less essential. The function of accommodation is the focusing of the rays of light, on the retina of each eye singly, which come from objects looked at within 20 feet from the eye; while the function of convergence is the turning of the eye so that the image of the object looked at will fall on corresponding parts of the retina in each eye. For a pair of normal eyes to view an object at a given distance the same amount of accommodation and convergence will be required. For instance, to view an object at 13 inches, it will be necessary to use three dioptries of accommodation and three meter angles of convergence, and if the object was brought nearer, the accommodation and convergence would increase an equal amount. The same nerve (third or motor oculi) supplies the muscles that perform both functions. Turning the eyes outward from their position of rest is called **negative convergence,** or abduction; turning the eyes inward from their position of rest is called **positive convergence,** or adduction. The number of degrees of convergence that can be used without changing the accommodation is called **relative convergence.** The total amount one has, the **Amplitude.**

onverge. Directed toward the same point.

onvex. That which has a rounded and elevated surface. The surface, if continued at the same radius of curvature, would form a complete circle, or sphere.

onvexo-concave. Convex on one side and concave on the other. If the convexity exceeds the concavity it is known as a periscopic convex lens. If the concavity exceeds the convexity it is known as a periscopic concave lens.

opiopia (kop-i-o'-pi-ah). (Gr. kopos = fatigue + ops = eye.) A worn-out state of the eye, caused by eye-strain.

oquille-Plano Lenses (plus 8D. on one side and minus 8D. on the other). MiCoquille are plus 4D. on one side and minus 4D. on the other. They are nearly always colored.

oreclisis (kor-ek'-lis-is). (Gr. kore = pupil + kleisis = closure.) That condition in which the pupil of the eye is obliterated.

orectasis (kor-ek'-tas-is). Dilatation of the pupil.

orectome (kor-ec'-to-me). An instrument used in cutting for iridectomy.

orectomy (ko-rek'-to-me). (Gr. kore = pupil + ektome = excision.) See Iridectomy.

orectopia (kor-ec-to'-pi-ah). (Gr. kore = pupil +

ektopos = out of place.) That condition in which the pupil is displaced.

Coredialysis (kor-e-di-al′-ys-is). An operation in which the iris is detached from the ciliary ligament for a new pupil.

Corelysis (ko-rel′-is-is). (Gr. kore = pupil + lysis = a loosening.) The loosening of adhesions between the capsule of the lens and the iris.

Coremorphosis (kor-e-mor′-pho-sis). (Gr. kore = pupil + morphosis = formation.) Creation of an artificial pupil.

Coreometer .(kor-e-om′-et-er). (Gr. kore = pupil + metron = measure.) A contrivance used for measuring the pupil.

Coreplasty (Gr. kore = pupil + plasso = I form). An operation for forming an artificial pupil.

Cornea (kor′-ne-ah). (L. corneus = horn-like.) The anterior one-sixth of the first tunic of the eyeball. It is transparent, convex, and fitted into the sclerotic like a watch crystal, having a radius of 7.8 mm. on its anterior surface and 6.5 mm. on the posterior surface. It has a vertical diameter of about 11 mm. and the horizontal diameter is 12 mm. In thickness it is about 1 mm. at the center, while at the periphery it is 1.12 mm. The cornea has no blood-vessels except for a narrow space about 1 mm. wide at the margin, derived from the anterior ciliary arteries. The venous roots become tributaries of the anterior ciliary vein. The cornea is well supplied with nerves and lymphatics. It serves to transmit light into the eye, and next to the layer of tears becomes the first refracting medium. It is convex in front and concave behind. Its curvature varies in different individuals. It is composed of five layers, arranged as follows, from without inward, namely: (1) Conjunctiva epithelium; (2) Bowman's membrane; (3) Cornea proper; (4) Membrane Descemet; (5) Endothelium. The first layer (conjunctiva epithelium) serves to protect the nerves in Bowman's membrane from cold, wind, and dust, and at the same time gives a highly polished surface to the cornea. The second layer (Bowman's membrane) is a layer of sensitive nerves and elastic tissue, and protects the cornea proper on the anterior side, and at the same time gives the cornea an elastic nature. The third layer (cornea proper) is the foundation layer of the cornea. It is composed of a horn-like substance and is non-sensitive and merely serves to keep the cornea in shape. The fourth layer (Membrane Descemet) is a layer similar to Bowman's membrane, and protects the cornea proper from any diseased condition from the posterior side. The fifth layer (endothelium) is a lining membrane which separates the aqueous humor from the fourth layer, and at the same time forms a sort of sac which contains the aqueous humor. The cornea has an index of refraction of 1.33. Its nerve supply arises from the ciliary nerves.

Cor'neal. Pertaining to the cornea.

Corneal Astigmatism. See Astigmatism.

Corneal Facets (făs'-ets). Small, plain, distinct surfaces of the cornea.

Corneitis (cor-ne-i'-tis). Inflammation of the cornea.

Corneo-iritis. Inflammation of the iris and cornea.

Corneosclera. The cornea and sclera taken together, forming the external coating of the eyeball.

Corradiation (kor-ra′-di-a′-shun). A conjunction or concentration of rays in one point.

Correction. Making good an abnormal condition, such as correcting an error of refraction.

Cortical (kor′-tik-al). (L. cortex = bark.) To be near the border. **Cortical Cataract** is that variety in which the opacity begins at the border of the crystalline lens and gradually spreads toward the center, which it sooner or later involves.

Cosecant (ko-se′-kant). The secant of the complement of an arc or angle.

Cosine (ko′-sin). The sine of the complement of an arc or angle.

Cotangent (ko-tan′-jent). The tangent of the complement of an arc or angle.

Couching. That condition in which the lens is displaced in cataract. This operation is now obsolete.

Coversed Sine (ko-verst′ sine). The versed sine of the complement of an arc or angle.

Cover Test. A test for muscular imbalance by covering one eye and observing its movement while uncovering, the point of fixation being established.

Cramp. A spasmodic muscular contraction.

Cribriform (krib′-ri-form). (L. cribrum = sieve + forma = form.) Perforated like a sieve.

Critical Angle (krit′-ik-al). The least angle of incidence at which a ray of light traveling in a denser medium is totally reflected at the surface which separates it from a rarer medium; also known as limit angle. The limit angle of crown glass is 40° 49′; that of flint glass, 37° 36′.

Crossed Diplopia. See Diplopia.

Crystalline (Gr. "crystal"). Clear, transparent. Resembling or of the nature of crystal.

Crystalline Lens (krys′-tal-lin). The lens of the eye, which resembles a crystal. It is a biconvexed, transparent, elastic body having a diameter of about 8.5 mm. and its axial thickness about 3.6 mm. It has a radius of curvature on the anterior surface of 10 mm., and that of the posterior surface 6 mm., while the eye is in a state of rest. It is located in the hyaloid fossa of the vitreous, just behind the pupil, and is made up of layers like an onion, which gives it an elastic nature. The lens itself is enclosed in the lens capsule, which is held in position by the suspensory ligaments. Its index of refraction is 1.43, and it represents from 16 to 19 diopteries of plus when the eye is at rest.

Crytometer. See Curtometer.

Cube (kub). (Gr. kubos = a die, a cube.) (a) A regular solid with six square faces; (b) to raise to the third power; (c) the third power of a number.

Cube Root. The cube root of a perfect third power is one of the three equal factors of that power. A number which has not a perfect third power has not three equal factors.

Cuneus (cu′-ne-us). (L. wedge.) The wedge-shaped portion of the occipital lobe of the cerebrum, situated between the occipital and calcarine fissures. The cuneus receives the cortical termini of the optic tract.

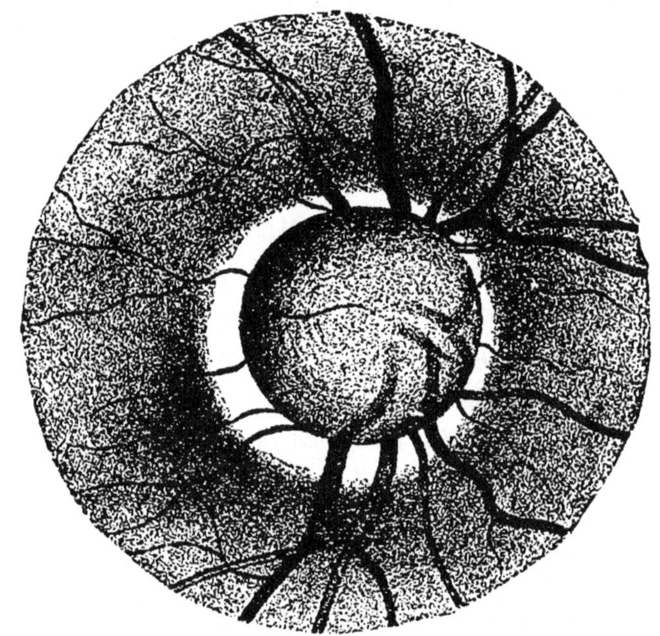

Cupped Disc.

Cupped Disc. That condition in which the optic disc has become cupped, as seen in glaucoma.

Curtom'eter. An instrument for measuring curved surfaces.

Curvature (curv'-a-ture). The bending of a line without forming angles.

Cutaneous (ku-ta'-ne-us). Pertaining to the skin.

Cyclitis (cyc-li'-tis). (Gr. kyklos = circle + itis = inflammation.) Inflammation of the ciliary body.

Cyclochoroiditis (si-klo-ko-roid-i'-tis). Inflammation of the choroid and ciliary body.

Cycloid (si-kloyd). Like a circle.

Cyclophoria (cyc-lo-fo'-ri-ah). (Gr. kyklos = circle

+ phora = movement.) That condition in which the vertical axis of the eye inclines to the right or left instead of standing vertically, the extrinsic muscles being at rest.

Cyclopia (si-klo′-pe-ah). (Gr. kyklos = circle + ops = eye.) A single eye in center of forehead.

Cycloplegia (cy-clo-ple′-gi-ah). (Gr. kyklos =-circle + plege = stroke.) Paralysis of the ciliary muscles.

Cycloplegic. A drug which produces paralysis of the ciliary muscles or muscles of accommodation

Cyclot′omy (Gr. kyklos = circle + tome = incision). Operation of cutting the ciliary body of the eye.

Cyclotropia (Gr. kyklos = circle + trope = turn). The actual turning of an eye on its optic axis.

Cylinder (cyl′-in-der). (Gr. kylindros = a roll.) See Lens.

Cylindrical. Relating to or the shape of a cylinder.

Cyst (sist). (Gr. kystis = bladder.) Any sac containing a liquid. Dermoid cyst is congenital. It is a painless, uninflammed spheroidal mass, situated generally at the outer angle of the orbit, on a level with the outer end of the eyebrow.

Cystitome (sis′-tit-om). An instrument used for opening the sac of the crystalline lens.

Cystot′omy (Gr. kystis = bladder + tome = incision). Incision of the capsule of the Crystalline Lens.

D. Abbreviation for dioptry, dexter, or dose.

Dacryadenalgia (dak-ry-ad-en-al′-gi-ah). (Gr. dakryon = tear + aden = gland + algos = pain.) Pain in a lacrimal gland.

Dacryagogue (dak′-ry-ag-og). (Gr. dakryon = tear + agogos = leader.) A medicine which causes a flow of tears.

Dacryoadenitis (dak-ry-o-ad-en-i′-tis). (Gr. dakryon = tear + aden = gland + itis.) Inflammation of a lacrimal gland.

Dacryocele (dak′-ry-o-cele). (Gr. dakryon = tear + kele = hernia.) A protrusion of the lacrimal sac.

Dacryocyst (dak′-ry-o-cyst). (Gr. dakryon = tear + kystis = sac.) The tear sac.

Dacryocystalgia (dak-ry-o-cyst-al′-gi-ah). (Gr. dakryon = tear + kystis = sac + algos = pain.) Pain in the lacrimal sac.

Dacryocystitis (dak-ry-o-cys-ti′-tis). Inflammation of the lacrimal sac.

Dac′ryoid. Resembling a tear.

Dacryoma (dak-ry-o′-ma). (Gr. dakryon = tear, and suffix oma, "morbid state.") A lacrimal tumor which causes an obstruction of the lacrimal puncta, so that the tears flow over the lids upon the cheek.

Dacryon (Gr. "a tear").

Dacryops (dak′-rė-ops). (Gr. dakryon = tear + ops = eye.) A watery eye. Applied to a swelling of the lacrimal sac or one of its ducts.

Dacryorrhea (Gr. dakryon = tear + rhoia = flow). Excessive or morbid flow of tears.

Dacryopyorrhea (Gr. dakryon = tear + pyon = pus + rhoia = flow). Discharge of pus from the lacrimal duct.

Dacryopyo′sis (Gr. dakryon = tear + pyosis = suppuration). A discharge of tears mixed with purulent matter.

Daltonism (dawl′-ton-izm). Color-blindness.

Day-blindness (day-blind′-ness). Partially blind by day, with better vision at night. See Hemeralopia.

Decameter (dek′-a-me-ter). Ten meters.

Decentered (de-cen′-terd) **Lens.** A lens with its optical center to one side or above or below the center.

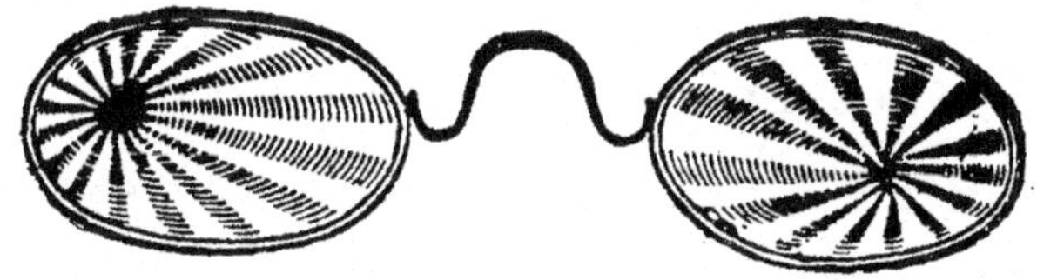

Decentered Lenses.

Decentering of Lenses. Instead of having a prism and a lens combined, where you wish to obtain the effect of both, it is possible to get the same result by simply decentering the optical center of the lens. The optical center of a plus lens is at the thickest part, and in the minus at its thinnest part, while the geometrical center of a lens is the point midway between all edges. A 1-dióptry lens decentered 10 mm. will give the effect of a 1° prism, while a 2-D. lens will only require to be decentered half this amount, or 5 mm.; a 3-D. lens, one-third of this amount, for the same effect, and so on according to the strength of the lens. To obtain the effect of a

2° prism these lenses must be decentered twice as much—that is to say, a 1-D. lens, 20 mm.; a 2-D. lens, 10 mm.; a 3-D., 6.3 mm. From this table one can easily figure the exact amount any lens should be decentered to obtain a given prismatic effect. **Law of Decentration:** Any lens is capable of producing as many prism dioptries as the lens possesses dioptries of re fraction, provided it is decentered 1 cm.—Prentice, Archives of Ophthalmology, Vol. XIX, No. 1 and No. 2, New York, 1890.

Decentration (de-cen-tra′-tion). The act of removing from a center.

Decimal (des′-i-mal). (L. decem = ten.) Pertaining to ten.

Decimeter (des′-i-me-ter). One-tenth of a meter.

Decomposition of Light. If parallel rays of sunlight pass through a prism it is not only refracted but it is also decomposed into its various colors. This is due to the unequal refrangibility of the different colored rays which form white light, the violet being refracted the most and the red the least, thus forming the spectrum.

Decussation (de-ku-sa′-shon). (L. decussare = to cross.) The act of crossing or intersecting; the crossing of two lines, rays, fibers of nerves, etc.

Through the decussation of the rays in the pupil of the eye the image of the object on the retina is inverted.

Defect (de-fect′). A departure from the normal. When speaking of defects of vision we mean the visual power of the eye is not normal.

Defining Power, Definition. The power of a lens to give a clear outline.

Denominator (de-nom′-i-na-tor). (L. denominare = to name.) (Arith.) That number placed below the line in fractions which shows into how many parts the unit is divided. (Alg.) That part of any expression under a fractional form which is situated below the horizontal line signifying division.

Deor′sumvergens (L. deorsum = downward + vergere = to incline). Downward turning of the eye.

Depilation (dep-il-a′-shun). (L. de = from + pilus = hair.) The removal or loss of the hair.

Deplumation (de-plu-ma′-shun). (L. de = from + pluma = feather.) Loss of eyelashes by disease.

Deprimens Oculi (dep′-ri-mens ok′-u-li). (L. deprimere = to depress.) The rectus inferior muscle.

Descemet's Membrane (des-ce-mets′ mem′-brane). The fourth layer of the cornea. See Cornea.

Descemetitis (des-em-e-ti′-tis). Inflammation of Descemet's Membrane.

Deviation (de-vi-a′-shun). (L. de = from + via = way.) Turning aside, as in strabismus. **Conjugate d.,** deviation of both eyes to the same side. **Minimum d.,** the smallest deviation of a ray that a given prism can produce.

Dexter, Dextra (dex′-ter, dex′-tra). On right side.

Dextrad (dex′-trad). (L. dexter = right.) Toward the right side.

Dextrophoria (deks′-tro-phor′-ia). (L. dexter = right + Gr. phoria = tending.) That condition in which the eyes turn to the right when the extrinsic muscles are in a state of rest.

Diacaustic (di′-a-kas′-tik). A curve formed by the consecutive intersections of rays of light refracted through a lens, and when reflected it is called catacaustic.

Diagonal (di-ag′-o-nal). (Gr. dia = through + gonia = corner, angle.) A line through the angles of a figure, but not lying in its sides or faces.

Diameter (Gr. diametros; dia = through + metron = measure). A straight line joining opposite points of a circle, drawn through the center.

Diaphaneity (di-af-a-ne′-i-ty). Transparency; the power of transmitting light.

Diaphanous (di-af′-a-nous). Having power to transmit rays of light, as glass.

Diaphragm (di′-a-fram). (Gr. diaphragma = a partition wall.) A term applied to the partition with a central aperture in optical instruments so that rays of light may be controlled. The iris with its pupil constitutes the diaphragm of the eye.

Diapyesis (di-ap-i-e′-sis). Suppuration.

Diffraction (dif-frak′-shun). (L. diffractus; diffringere = to break up.) Deflection or decomposition of light in passing by the edges of opaque bodies or through small apertures.

Diffusion (dif-fu′-shun). n. (Optics) A spreading or scattering of rays of light, causing a blurred image by imperfect refraction.

Digit (dij′-it). (L. digitus = finger.) The number represented by any one of the ten symbols 0, 1,

2,. ..9. The term is more often used to designate one of the ten symbols mentioned.

Dilatant. A medicine that causes dilatation.

Dilatation (di-la-ta′-shun). (L. dilatare = to expand.) The expansion of any orifice or canal.

Dilator (di-la′-tor). Dilator iris refers to the radiating fiber of the iris which dilates the pupil.

Diopter (di-op′-ter). (Gr. dia = through + opsomai = I shall see.) A leveling instrument of ancient times, equipped with sights at both ends and a water level in its center. The theodolite, by Hipparchus, the ancient Greek mathematician.

Dioptometer (di-op-tom′-e-ter). An instrument for testing ocular refraction.

Dioptometry (di-op-tom′-e-tre). The measurement of ocular accommodation and refraction.

Dioptral, a. Applied to the refractive power of ophthalmic lenses numbered according to the metric system, and in which the unit of power has a focal length of 1 meter. Thus, a 1-dioptry lens is specifically a member of the dioptral system, whereas a 40-inch telescope lens is a member of a dioptric system.

Dioptric (di-op′-tric). See Dioptry.

Dioptrical (di-op′-tri-kal). Of or pertaining to dioptric.

Dioptrics (di-op′-trics). That branch of optics which treats of the refraction of light by any transparent media, as air, water, or glass.

Dioptron (di-op′-tron). (Surg.) A dilating speculum.

Dioptry, n. (di-op′-tri). (Gr. dia = through + opsomai = I shall see.) The unit for expressing the refractive power of a lens. The refractive power of a lens which will focus parallel rays of light at a distance of 1 meter. A lens of 2 dioptries (2 D.) has a focal length of ½ meter. Synonyms: Dioptre, Dioptric. See Focus

Diplocoria (dip-lo-ko′-re-ah). (Gr. diplous = double + kore = pupil.) Double pupil.

Diplopia (dip-lo′-pe-ah). (Gr. diplous = double + opsis = vision.) Double vision; seeing one object as two. The object of convergence is to direct the yellow spot (or macula lutea) in each eye toward the same point, so as to obtain single vision; diplopia, or double vision, at once resulting when the image of an object falls on parts of the retina which do not exactly correspond in the two eyes.

Binocular d., double vision only when the two eyes are exposed to view. **Heteronymous d.**, where the object seen with the right eye appears on the left side, and that of the left eye on the right side. **Homonymous d.**, where the object of the right eye appears on the right side and the object of the left eye on the left side. **Monocular d.**, diplopia with a single eye.

Diplopiometer (dip-lo-pi-om′-e-ter). An instrument for measuring diplopia.

Disc (disk). A round body which resembles a small circular plate. **Optic d.**, a whitish circular spot in the retina representing the entrance of the optic nerve into the globe of the eye.

Discission (dis-ish′-un). (L. discindere = to split.) The rupture of the capsule of the crystalline lens in the operation for soft cataract.

Diseases of the Eye. The diseases of the eye are

many, but nearly all of them can be directly or indirectly attributed to eye-strain or impurity of the blood. First, relieve any eye-strain by glasses; second, keep the bowels regular; third, fresh air and exercise. When the patient requires further attention, proper treatment should be instituted.

Disparate Points (dis′-par-at). (L. disparare = to separate.) Points on the two retinae upon which light does not produce the same impression.

Disperse (L. dispersus; dispergere = to scatter about).

Dispersing Lens (dis-per′-sing). Same as concave lens.

Dispersion (dis-per′-shun). The process of scattering the rays of light through any kind of a lens.

Distichiasis, Distichia (dis-te-ki′-a-sis, dis-tik′-e-ah). (Gr. di = double + stichos = row.) That condition of the eyelashes in which a second row rubs against the cornea, causing inflammation.

Divergence (di-ver′-gens). (L. di = apart + vergere = to incline.) To turn outward from parallelism.

Dividend (div′-i-dend). (L. dividere = to divide.) A number or quantity to be divided by another is called the dividend.

Donders (Frans Cornelis). A Dutch physician, born at Tilburg, Holland, May 27, 1818. He was educated at Utrecht, where he became a professor of physiology.

Double Prism. An opaque disc found in test-case with a slit-like opening. Over this slit there are two prisms with their bases together. Used for testing for muscular imbalance. To test for muscular imbalance use a small round bright frosted light (about ⅛ to ¼ inch, at 20 feet), the double prism will make this light appear as two, both of which will be displaced from their true position. Over the other eye place the red glass disc which will color the light red. The patient will then see three lights, two white, one red. If the red light is seen midway between the two white, no muscular imbalance is present; if not, the prism that will bring it there is the measure of deviation. The two prisms used in this disc are of five diopters each, joined base to base. When the line formed by the bases is placed horizontally before the pupil of an eye the light will be seen as two, separated vertically from their true position leaving the other eye, which is wearing the red glass disc, to see the light in its true position.

Double Vision. Seeing one object as two. See Diplopia.

Doublet (doub'-let). Composed of two lenses.

Duct (dukt). A tube for conveying a fluid.

Dura Mater. The outermost membrane of the brain, spinal cord, optic nerve, and capsule of Tenon.

Dural Sheath. The external covering of the optic nerve.

Dynameter (dy-nam'-e-ter). An instrument for determining the magnifying power of telescopes.

Dynamic (Gr. dynamis = power). The powers whereby bodies are put in motion.

Dynamic Refraction (dy-nam'-ic). (Gr. dynamis = power.) The refraction of the eye (dioptric power) while using all of its accommodation and adjusted for the near point of vision. The difference between the dynamic and static refraction is known as the Amplitude of Accommodation.

Dynamic Skiametry (di-nam′-ic sky-am′-e-try). (Gr. dynamis = power + skia = shadow + metron = measure.) Is defined as the measuring of the refraction of an eye by the shadow test (retinoscopy) while the accommodation is under tension. It is the opposite of static skiametry, where the accommodation is relaxed.

The value of dynamic skiametry is said to be in its offering a mechanical means for absorbing spasms of accommodation, and thus revealing latent errors without recourse to cycloplegics. In the practice of the dynamic method the patient is required to accurately fix his vision on a test card that is situated the same distance away (say, 16 inches) that the nodal point of an examiner's eye (behind the peephole of his skiascope) is situated during the examination; that is, the **fixation** and **observation** must be exactly the same. Under these conditions, if the examiner will keep adding **all** the plus spherical lens power possible before reversal of the shadow takes place (even though the shadow does not indicate hyperopia) it will be found that an eye will surrender that portion of its accommodation which is in **excess** of what is needed to correctly harmonize with the convergence required for the distance at which the examination is made. With the dynamic method no allowance in strength of lens is made for the working distance, as by the static method, and where the patient is presbyopic or has latent hyperopia there will be a difference in the findings between the dynamic and static.

The recognized text-book on the subject is called "Dynamic Skiametry in Theory and Practice," by Andrew Jay Cross, D.O.S., of Columbia University, New York City, who is the originator of the method.

Dynamometer (dy-na-mom'-e-ter). An instrument for measuring force or power; especially the muscular power.

Dynamometry (dy'-na-mom'-e-try). The process of measuring force while doing work.

Dyslexia (dis-lex'-se-ah). (Gr. dys = bad + lexis = word.) Inability to read caused by a disease of the brain. Vision is good, but the power to read is wanting.

Dysopsy (dys-op'-sy). (Gr. dys = faulty + opsis = vision.) Dimness of vision.

ECCENTRIC (ek-sen'-trik). (Gr. ek = out + kentron = center.) Away from a center.

Ecchymosis (Gr. ek = out + chymos = juice). An extravasation of blood into tissue.

Ectasia (ek-ta'-se-ah). Abnormal distention or dilatation of a part.

Ectiris (ek-ti'-ris). (Gr. ektos = outside + iris.) The external portion of the iris.

Ectochoroidea (ek-to-cho-roi'-de-ah). (Gr. ektos = outside.) The outer layer of the choroid coat.

Ectocornea (ek-to-kor'-ne-ah). Outer layer of the cornea.

Ectoretina (ek-to-ret′-in-ah). (Gr. ektos = outside.) Outermost layer of the retina.

Ectropion (ek-tro′-pi-on) **Ectropium** (Gr. ek = out + trope = a turning). Turning out or inside out of the edge of an eyelid.

Edema (e-de′-mah). (Gr. oidema = a swelling.) An accumulation of serum in the cellular tissue.

Efferent (ef′-er-ent). (L. effere = to bring out.) Conveying outward, as from center to periphery; applied to motor nerves and vessels conveying from the center. The opposite of afferent.

Em′bolism (Gr. embolisma = a patch). Obstruction of a vessel by an embolus.

Em′bolus. A clot or plug which obstructs a blood-vessel.

Emergent ("to come out of"). A ray of light after having passed through a refracting medium.

Emissive. Radiating.

Emmetropia (em-met-ro′-pi-ah). (Gr. emmetros = in measure + ops = eye.) That condition in

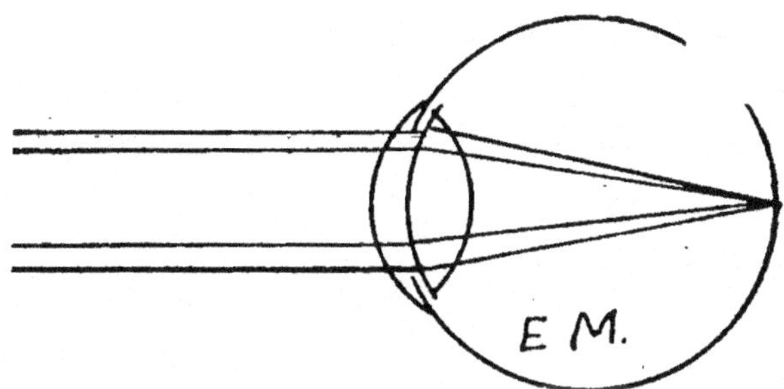

An emmetropic eye receiving one set of parallel rays. It must be remembered that the three rays representing the set come from one point, but the point is so far away that the rays appear to be parallel because the divergence is so slight.

which all parallel rays of light after entering the eye are brought to a focus on its retina, while the muscles of accommodation are in a state of rest.

Emmetropia has no reference whatever to sight, or disease, but simply means that the optical apparatus has the correct focal length for the globe of the eye. In other words, no lens is required for distant vision. When this condition does not exist the eye is out of measure, or ametropic.

Emphyse'ma (Gr. en = in + physema = a blowing). The infiltration of air into the cellular tissues of the orbit. May be caused by rupture of the lachrymal sac.

Encan'this (Gr. en = in + kanthos = canthus). A minute tumor in the inner canthus of the eye.

Endothelium (Gr. endon = within + thele = nipple). A layer of flat cells lining serous cavities, blood-vessels, and lymphatics. The fifth layer of the cornea.

Energy. (Physics) Capacity for performing work.

Enervate (en-er-vate). To deprive of nerve, force, or strength; to render feeble.

Enophthalmus (en-of-thal'-mus). (Gr. en = in + ophthalmos = eye.) A condition where the eyes are deep-seated.

Enstrophe (en'-stro-fe). (Gr. en = in + strophe = a turning). A turning inward.

En'tad (Gr. entos = within). Toward a center.

Entochoroidea (en-to-cho-roi'-de-ah). (Gr. entos = within + chorioeides = choroid). The inner layer of the choroid.

Entocornea (en-to-cor'-ne-ah). (Gr. entos = within.) Descemet's membrane.

Entoptic (en-top′-tic). (Gr. entos = within + optikos = visual.) Situated within the eye.

Entoptic Phenomenon (en-top′-tic phe-nom′-e-non). That which is peculiar with itself, such as Muscae Volitantes.

Entoptoscopy (en-top-tos′-co-py). (Gr. entos = within + optos = visible + skopeo = I view.) Inspection of the interior of the eye.

Entoretina (en-to-ret′-in-ah). The nervous or inner layer of the retina.

Entropion (Gr. en = in + trope = a turning). See Entropium.

Entropium (en-tro′-pi-um). A turning in or inversion of the eyelid or eyelashes.

Enucleate (e-nu′-cle-ate). (L. e = from + nucleus = kernel.) To remove from its cover.

Enucleation (e-nu′-cle-a′-shun). Operation for the removal of the eye.

Ephidro′sis. (Gr. "I sweat.") An excessive secretion of the sweat glands of the eyelids. It causes itching, irritation, and inflammation of the skin and conjunctiva. It is difficult to cure.

Epicanthus (ep-i-can′-thus). (Gr. epi = upon + kanthos = canthus.) A fold of skin projected over the inner canthus.

Epiphora (e-pif′-or-a). (Gr. epi = upon + phoria = tending.) An overflow of tears, causing them to run down the cheek.

Episclera (Gr. epi = upon + skleros = hard). The connective tissue between the sclera and the conjunctiva.

Episcleral (ep-i-scle′-ral). Situated over the sclera of the eye.

Episcleritis (ep-i-scle-ri'-tis). Inflammation of the outer layers of the sclera.

Epithelio'ma. Cancer composed largely of epithelial cells, and is the most frequent of malignant growths affecting the eyelid. It seldom appears before the age of forty.

Epithelium (ep-i-the'-le-um). (Gr. epi = upon + thele = nipple.) The non-vascular, external layer of the skin and mucous membrane. First layer of the cornea.

Equal (e'-kwal). (L. aequalis = equal.) Having the same value.

Equation (e-kwa'-shun). A proposition asserting the equality of two quantities and expressed by the sign "=" between them. In Algebra, an equality which exists only for particular values of certain letters called unknown quantities.

Equilateral (e-kwi-lat'-e-ral). (L. aequus = equal + latus = side.) Having all of the sides equal.

Equil'ibrating Operation. Tenotomy of the muscle, which antagonizes a paralyzed muscle of the eye.

Errors of Refraction. Abnormal conditions of refraction in the eye.

Erythropsia (erythro'-psia). (Gr. erythros = red + opsis = vision.) Red vision; a condition in which all objects appear to be tinged with red.

Eserine (es'-er-een). An alkaloid obtained from the calabar-bean, which will cause contraction of the pupil. It has an action exactly opposite to that of atropine, since it places the iris and ciliary muscle in a state of tonic contraction. Consequently, miosis develops, so that the pupil is about the size of a pin's head, with adjustment of the eye for the near point, as if marked myopia were present. We generally apply sulphate of eserine in 1 per cent solution. This solution, when freshly prepared, is colorless, but after some days becomes red, although without losing its activity. The instillation of eserine produces, simultaneously with the changes in the iris, a feeling of great tension in the eye, and frequently headache, and even nausea, so that with many persons it cannot be employed. For this reason, hydrochloride of pilocarpine, prescribed in a 1 to 2 per cent solution, is recommended as a miotic for ordinary use. Its solution keeps better than that of eserine, and does not act as powerfully as the latter, but is not accompanied by any unpleasant complication. Eserine is best reserved for those cases in which pilocarpine is ineffectual.

Esophoria (es-o-fo′-ri-ah). (Gr. eso = inward + phoria = a tending.) That condition of the eyes in which the visual axes, although parallel when in use for distant vision, deviate inward when the extrinsic muscles are in a state of rest.

Esotropia (e-so-tro′-pi-ah). (Gr. eso = inward + trope = turn.) This term expresses a stronger meaning than Esophoria, in which there is merely a tendency, while in Esotropia there is a positive and visible appearance of the eyes turning inward.

Evolution (ev-o-lu′-shun). (L. evolvere = to unroll.) The extraction of roots from powers.

Excavation (ex-cav-a′-shun). (L. excavare = to hollow out.) Excavation of optic nerve; cupping or hollowing of the optic disc.

Exophoria (ex-o-fo′-ri-ah). (Gr. exo = outward + phora = a tending.) That condition of the eyes in which the visual axes, although parallel when in use for distant vision, deviate outward when the extrinsic muscles are in a state of rest. The internal rectus muscles are overworked from this continual convergence and relieved only when the lids are closed, or prisms worn, base in, that are strong enough to bend the light to suit the eyes in their deviating position of rest.

Excphthalmic Goiter (eks-off-thal′-mik goi′-ter). A goiter with exophthalmos and cardiac palpitation; Basedow's disease; Graves' disease. The most prominent symptoms are protrusion of the eye, excited action of the heart, enlarged thyroid (goiter), and certain nervous phenomena. The protrusion is almost invariably bilateral, though not infrequently greater on the right side. The upper lids do not follow the eyeball in looking down (Von Graefe's sign); infrequency of involuntary winking (Stellwag's sign) and abnormal width of the palpebral aperture are also found.

Exophthalmos (ex-of-thal′-mos). (Gr. ex = out + ophthalmos = eye.) Abnormal protrusion of the eye.

Exor′bitism (L. ex = out + orbita = orbit). Protrusion of the eyeball.

Exotropia (ex-o-tro′-pi-ah). (Gr. exo = outward + trope = turn.) When the eye is turned outward from parallelism. See Divergent Strabismus.

Extraction (ex-trak′-shun). (L. extrahere = to

draw out.) The removal of a body by surgical means.

Extravasation (eks-trah-vas-a′-shun). (L. extra = out of + vas = vessel.) The escape of fluids from their proper vessels, into surrounding tissues.

Extremes (eks-tremz′). (L. extremus = outermost.) The first and last terms of a proportion or of any other related series of terms.

Extrin′sic. Of exterior origin. E. Muscles are those on the outside of the organ.

Eye (L. oculus = eye). The organ of sight. The function of each eye, taken singly, is to form upon the retina, or nervous membrane which

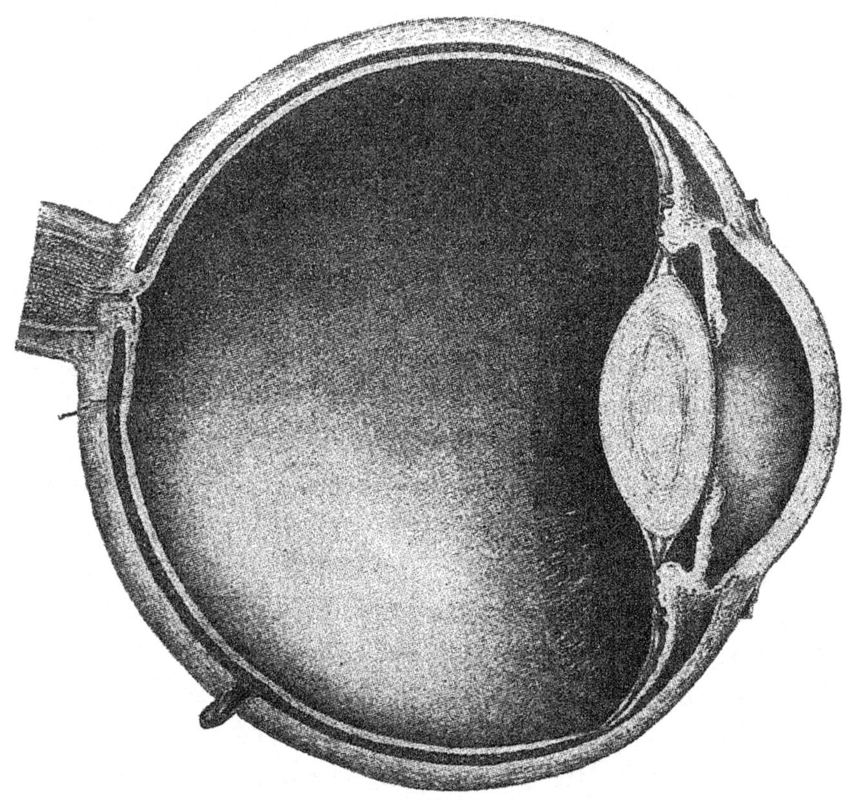

Eye.

lines the inside and back part of the organ, a sharply defined inverted image of any object looked at. The eye resembles a photographer's camera inasmuch as the image produced upon the retina is precisely the same as that produced on the ground glass of a camera. By means of the optic nerve the image that is received on the retina is conveyed to the brain, which recognizes the visual appearances and completes the act of seeing. More than this we do not know, but we do know that it depends upon the sharpness and clearness of the retinal image. If the image is blurred and indistinct it will be impossible for the brain to recognize the object accurately.

Eyebrows. They are two projecting arches of integument covered with short, thick hairs, which form the upper boundaries of the orbits.

Eye Ground. The inside and back part of the eye. The Fundus.

Eyelashes. The hair of the eyelids.

Eyelids. The anterior covering of the eye; that portion of movable skin with which the eyeball is covered or uncovered at will, protecting it from injury by their closure. The upper lid is the larger, the more movable of the two, and is supplied by a separate muscle, levator palpebrae superioris. When the eyelids are open an elliptical space is left between their margins, the extremities of which correspond to the junction of the upper and lower lids, and are called canthi. The outer canthus is more acute than the inner, and the lids here lie in close contact with the globe, but the inner canthus is prolonged for a short distance inward, toward the nose, and the two lids are separated by a triangular space, the lacus lachrymalis. At the commencement of the lacus lachrymalis and on the margin of each eyelid is a small conical elevation, the lachrymal papilla (the puncta), the apex of which is pierced by a small orifice, the commencement of the lachrymal canal. **Structures of the Eyelids:** The eyelids are composed of the following structures, taken in their order from without inward: Integument, areolar tissue, fibers of the orbicularis muscle, tarsal cartilage, fibrous membrane, meibomian glands, and conjunctiva. The upper lid has, in addition, the aponeurosis of the levator palpebrae. The integument is extremely thin, and continuous

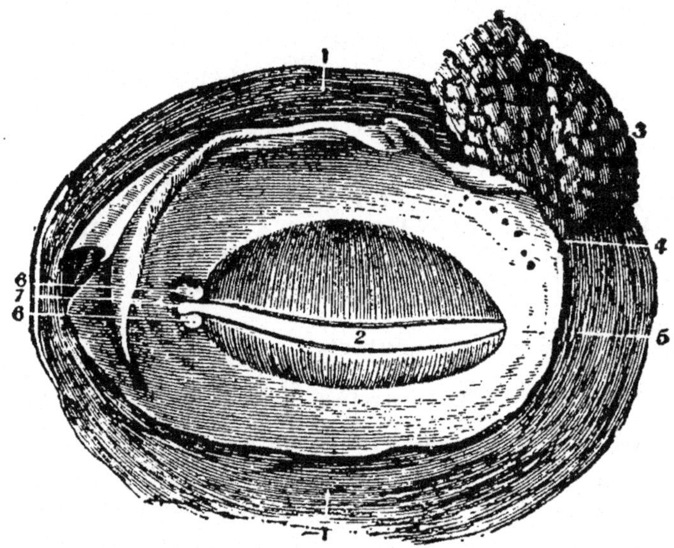

POSTERIOR VIEW OF EYELID SHOWING HOW THE TEARS ENTER THE CONJUNCTIVA.

1. Orbicularis Palpebrarum Muscle.
2. Opening between the lids (Palpebral Fissure.)
3. Lachrymal Glands, where the tears have their origin
4. Its ducts opening in the fold of the Conjunctiva.
5. Conjunctiva lining inside of lid.
6. Puncta Lacrimalia, through which the tears pass.
7. Inner Canthus.

at the margin of the lids with the conjunctiva. **The Subcutaneous Areolar Tissue** is very lax and delicate, seldom contains any fat, and is extremely liable to serous infiltration.

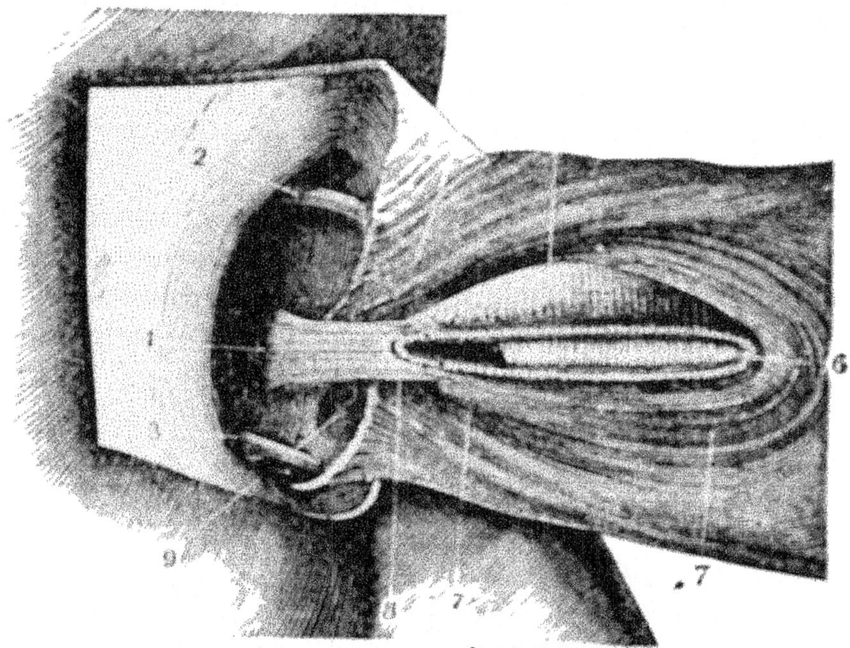

POSTERIOR VIEW OF THE PALPEBRAL (EYELID) WITH THE CONJUNCTIVA REMOVED

1 Origin of the Tensor-tarsi Muscle
2 Superior Oblique Muscle after passing through its Trochlea
3. Inferior Oblique Muscle
4 Attachment of Orbicularis Palpebrarum on Nasal side.
5 Tarsal Cartilages showing position of Meibomian Glands
6 Opening between the lids known as the Palpebral Fissure
7 Lower part of Orbicularis Palpebrarum Muscle
8 The Insertion of the Tensor-tarsi Muscle near the Puncta
9 Lachrymal Sac in the nose

Eyepiece. The lens or combination of lenses at the eye end of a telescope or other optical instrument, through which the image formed by the object glass is viewed.

Eyesight. The sense of seeing; sight of the eye; viewing; observation.

FACTOR (fak'-tor). (L. facere = to do.) One of two or more numbers which when multiplied together produce a given number.

Facultative (fak'-ul-ta-tiv). (L. facultas = faculty.) The power or ability to maintain extra effort whenever called upon.

Falling Eyelashes. (See Milphae and Madarosis.)

False Image. The image seen with the deviating eye.

False Myopia. Due to a spasm of accommodation, where the crystalline lens is kept convexed by the spasm and simulates true myopia.

Far Point. The far point or punctum remotum is the most distant point at which an object may be seen clearly, with the muscles of accommodation at rest. Properly speaking, the far point is an optical and not a visual point, and is that point from which rays of light will focus on the retina, the eye being in a state of rest.

Fascia (fash'-e-ah). A band or sheet of tissue connecting and investing muscles.

Field of Vision. The area or space which the fixed eye can see.

Filtration Angle. (See Iritic Angle.)

Fissure (fish'-ur). (L. findo = to split.) In anatomy, a cleft, or slit. **Palpebral fissure** is the opening between the margins of the eyelids. **Sphenoidal fissure** is a large split-like opening situated in the upper and back part of the orbit. **Spheno-maxillary fissure** is an opening in the back part of the orbit between the sphenoid, maxillary and palate and malar bones. **Cho-**

roidal fissure is the opening in the choroid through which the optic nerve passes to form the retina.

Flap Extraction. Removal of cataract by making a flap in the cornea.

Floating Specks. Small floating opacities in the humors of the eye. (See Muscæ Volitantes.)

Focal (fo′-kal). Pertaining to a focus. **F. Depth,** penetrating power of a lens. **F. Distance,** distance between the center of lens and its principal focus.

Focal Length of Lenses, in inches, centimeters and millimeters taken from the basis of forty inches as the equivalent to one meter.

Dioptries	English Inches	Centimeters	Millimeters
12	333	833	8333
25	160	400	4000
37	108	270	2703
50	80	200	2000
62	64½	161	1613
75	53	133	1333
.87	46	115	1150
1.00	40	100	1000
1.12	36	89	893
1.25	32	80	800
1.37	29	73	730
1.50	27	67	667
1.62	25	62	617
1.75	23	57	571
1.87	21	54	535
2.00	20	50	500
2.25	18	44	444
2.50	16	40	400
2.75	15	36	364

Dioptries	English Inches	Centimeters	Millimeters
3.00	13	33	333
3.25	12	31	308
3.50	11	29	286
3.75	10½	27	267
4.00	10	25	250
4.50	9	22	222
5.00	8	20	200
5.50	7	18	182
6.00	6½	17	167
6.50	6	15	154
7.00	5½	14	143
7.50	5¼	13	133
8.00	5	12½	125
9.00	4½	11	111
10.00	4	10	100
11.00	3½	9	91
12.00	3¼	8	83
13.00	3	7½	77
14.00	2¾	7	71
15.00	2⅔	6⅔	66⅔
16.00	2½	6¼	62½
18.00	2¼	5½	55½
20.00	2	5.	50

The above table is approximately correct, yet there is a slight difference in close figuring, but is correct as far as the optometrist is concerned; for instance, a + 1-D. lens has a focal length of 39.37 inches, while we call it 40.

Focal Planes. Straight lines through the foci perpendicular to the principal axis.

Focus (fo′-kus). The point produced by light coming to or going from a point. **First Principal Focus** is at the point the light leaves as divergent rays and emerges from the optical system as parallel to the principal axis. **The Second Principal Focus** is the point where the emergent rays cross each other when the incident rays have been parallel to the principal axis. **Negative Focus** is the point from which rays of light appear, to, but do not come from, the focus of a minus lens. **Secondary Focus.** Any focus of the secondary axis.

Fogging System. The system of fitting glasses by first making the patient artificially myopic by means of plus spheres, if they are not already myopic, the idea being to relax all accommodation before using cylinders.

Folders. A term employed for eye-glasses that can be folded up and placed in a small pocket.

Follicle (fol'-ik-l). (L. folliculus = a small bag.) A small secretory cavity or sac.

Follicular (fol-ik'-u-lar). Containing follicles. **F. Conjunctivitis.** A form of conjunctivitis marked by the presence of follicles. This occurs generally in children, and is characterized by the formation of small, clear elevations, consisting of adenoid tissue, in the conjunctiva of the lower lid; in some cases they are present also in the retrotarsal fold of the upper lid.

Fontana, Spaces of. In the anterior chamber of the eye, where the corneal margin joins the base of iris and sclerotic, we find a number of delicate bands of tissue passing from the membrane of Descemet to the base of the iris. These are known as the ligamentum pectinatum iridis and between them small comb-like openings leading into Schlemm's Canal, known as the Spaces of Fontana.

Foramen (fo-ray'-men). (L. foro = to bore a hole.) A hole or opening through any bone or a membranous structure. **Infraorbital F.**, the external opening of the infraorbital canal, on the anterior surface of the body of the maxilla; **optic F.**, the opening between the lesser wing and body of the sphenoid transmitting the optic nerve and ophthalmic artery. (For other foramen, see under orbit.)

Force. (L. fortis = strong.) (Physics.) Any action between two bodies which changes, or tends to change, their relative condition as to rest or motion; or, more generally, which changes or, tends to change, any physical relation between them, whether mechanical, chemical, or any other kind; as, the force of gravity.

Fornix. (L. arch, vault.) A vault-like space.

Fornix Conjunctiva. The turn or fold of the conjunctiva.

Fossae Patellaris (pa-tel-la'-ris) (meaning dish-like depression). The depression in the anterior surface of the vitreous body in which the crystalline lens lies. Also called the Hyaloid Fossa.

Fossa. (L. a ditch.) A pit, cavity or depression.

Fovea (foh'-ve-ah). (L. fodio = to dig.) A small depression. **F. Centralis** is employed to designate the little depression in the center of the macula lutea.

Fraction (frak'-shun). (L. frangere = to break.) One or more of the equal parts of a unit.

Frame Fitting. There are times when patients complain that their glasses are not comfortable, yet they have the right correction. The cause of the trouble is sometimes found in the improper adjustment of the frames. The fitting of a frame is very important, and if neglected will sometimes destroy the benefit of the most carefully fitted lenses. When a student understands the relation between accommodation and convergence the value of frame fitting becomes easily understood. A convex lens, with its curved surfaces, may be described as made up of an infinite number of prisms with their bases meeting at the center; a concave lens, in a like manner, is made up of an infinite number

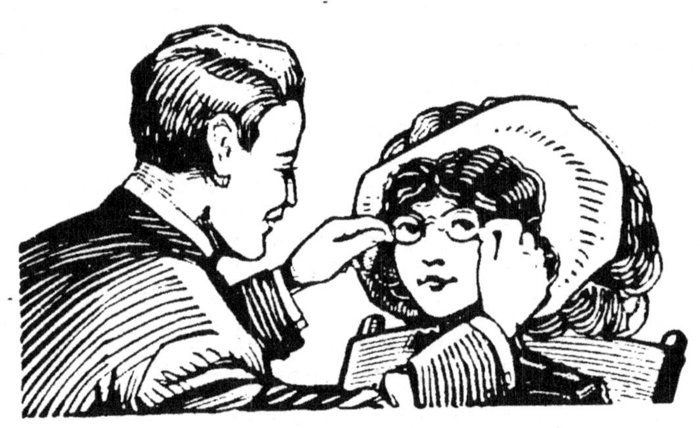

Frame Fitting.

of prisms with their bases outward. When a person looks through the inner side of a convex lens, as he is compelled to do when the frames are too wide for the pupillary distance, he is looking not only through convex lenses, but also through prisms with their bases outward; when the frames are too narrow he looks through prisms with their bases inward. With concave lenses, of course, this condition will be reversed, and besides giving a prismatic effect, will cause the unbalancing of accommodation and convergence.

The subject of frame fitting has always been and always will be more or less of a problem to the student, but after a little practice and careful attention it becomes a very easy matter. I will here mention a few points which may be of assistance to my fellow-student:

1st. See that the pupillary distance is correct and that the patient is looking through the center of lenses. If glasses are to be worn constantly it is best for the adjuster to stand off, say, about three feet, and direct the patient to look between his eyes, so adjusting frames that the patient will be looking through the centers of lenses. For reading glasses the optical centers should be slightly closer and lower, and the top of the lenses must be inclined forward, so as to be as near as possible at right angles to the line of vision. In this way better vision is enjoyed.

2d. The lenses should be placed as near the eye as the lashes will permit.

3d. Never prescribe a small lens for a large face nor a large lens for a small face, but always make the lenses as large as you possibly can without interfering with the patient's appearance, and at the same time see that the pupillary distance is correct. In the fitting of spectacles see that the angle of crest saddles the nose nicely, and that the temples are long enough to go around the ear without showing underneath. See that the temples are not too far from the face and at the same time do not press on the flesh. If you desire to tilt the lenses do not bend temples, but bend the end piece. All glasses should tilt outward from the top, but reading glasses more than distant ones. Cylinders should always be worn as spectacles, as it is very important that they should be held in their correct position.

It is always best for one who is just commencing to practice to supply himself with a full set of measuring frames. They are put up and sold by all wholesale optical houses. The optical houses also supply cards on which are printed the various dimensions. Then by finding a sample frame among your set that about fits your patient you lay it down on the card, allowing for any change you wish to make, and you can easily figure the exact dimensions.

Function (funk′-shun). (L. functio = to execute.) The special duties which an organ or group of organs has to perform. (Math.) A quantity so connected with another quantity that if any alteration be made in the latter there will be a consequent alteration in the former. Each quantity is said to be a function of the other. Thus, the circumference of a circle is a function of the diameter.

Fundus (fun′-dus). That portion of a hollow organ farthest from the entrance. The fundus of the eye is seen by means of the ophthalmoscope, namely, the retina, blood vessels, choroid, optic disc, collectively.

Fuscin (fus′-sin). (L. fuscus = dusky.) A brown pigment of the retinal epithelium.

GANGLION (gang'-gle-on). (Gr. "a knot.") In anatomy, a knot-like aggregation of nerve cells. It is a partly independent nerve center, with distinct functions in connection with nearby structures. **Ciliary G.,** sometimes called ophthalmic or lenticular, is about the size of a pin's head, situated in the back part of the orbit, between the external rectus muscle and the optic nerve. The three nerves which enter it are, one from the nasal branch of the ophthalmic (sensory), one from a branch of the third nerve (motor), and a root of the sympathetic. From it passes off about ten filaments which pierce the posterior part of the sclera supplying the ciliary muscles, the iris, and the cornea. **Gasserian** or **Semilunar G.** lies in a depression (cavum Meckelii) on the anterior surface of the petrous portion of the temporal bone near the apex. It is a flat expansion on the sensory root of the (fifth) trigeminal nerve, receiving on its inner side filaments from the carotid plexus of the sympathetic and giving off the ophthalmic, superior maxillary and inferior maxillary. The ophthalmic nerve is a sensor nerve. It supplies sensation to cornea, ciliary muscles, iris, lachrimal gland, mucous membrane of nose, skin of eyelids, eyebrows, forehead and nose. Just before entering the orbit, through the sphenoidal fissure, it divides into three branches, lachrimal, frontal and nasal.

Geometrical Center. A point midway between all edges.

Geometry. (Gr. geometria = to measure.) That

branch of mathematics which investigates the relations, properties and measurement of solids, surfaces, lines and angles; the science which treats of the properties and relations of magnitudes; the science and relations of space.

Generic Compounds. Lenses having spherical and cylindrical curvatures of the same species; that is, both convex or both concave. Contrageneric compounds have one surface convex, the other concave.

Glabel'la, Glabel'lum. (L. glaber = smooth.) Space between the eyebrows.

Gland. (L. glans = acorn.) The name applied to organs which separate from the blood any fluid whatever.

Bruch's Glands. The lymph-follicles of the conjunctiva. **Henle's Glands** are a number of follicular cavities formed by irregular folds in the epithelium of the tarsal conjunctiva.

Krause's Glands, the tubular glands which lie at the border of the tarsi near the fornix. These are regarded as accessory glands. **Ciliary Glands** are the sweat glands of the eyelids and located in several rows close to the free margin of the lid. They are also known as the **Glands of Moll.** **Lachrimal Glands,** the glands which secrete the tears. They are located in a depression of the frontal bone at the upper and outer angle of the orbits. The gland is divided into two parts, the superior and inferior and attached to the bony roof of the orbit by the tarso-orbital fascia. The ducts, about ten in number, open into the fornix conjunctiva. Its nerve supply is the smallest of the three branches of the ophthalmic and known as the lachrimal nerve. **Meibomian Glands** (see Meibomian). **Tarsal Glands** (same as Meibomian).

Glass. A hard, brittle, artificial substance formed by the fusion of silica, potash and lead. Under the best conditions it is quite transparent. Nothing definitely is known as to its origin. The Egyptians used it, and glass has been discovered amongst the ruins of Pompeii.

The media out of which lenses are made. Crown glass for optical lenses, sometimes combined with flint glass.

Glaucoma (glau-ko′-mah). (Gr. glaukos = greenish gray.) A disease of the eye characterized by increased intraocular tension. In order to fully understand this disease it will be necessary to study thoroughly the anatomy of the eye, and in doing so pay particular attention to Schlemm's Canal and the Spaces of Fontana, situated in the first tunic between the sclerotic and cornea. These canals are said to carry away the excess of aqueous humor. The theory most generally accepted is, that the vitreous humor is formed in the choroid and ciliary body and passes through the hyaloid membrane into the vitreous cavity; from there it filters through the suspensory ligaments into the posterior chamber, where it becomes watery, and is known as the aqueous humor. After passing through the pupil into the anterior chamber it is said to pass through the Spaces of Fontana into Schlemm's Canal. In this way one can readily see that if the iris was attached to the lens, as it is in cases of iritis, or by the straining of the ciliary muscles, as in hypermetropia, thus closing the Spaces of Fontana, the drainage system would be blocked, while the humors continue forming, resulting in a painful intraocular pressure. Glaucoma may be divided into two kinds, primary and secondary. Primary, when it makes its appearance in a healthy eye, or with a disease like cataract. Secondary, when caused by a disease like iritis. It is a progressive disease, and unless checked by treatment ends in permanent blindness.

Symptoms of Glaucoma. (1) Pain, sometimes of a neuralgic character. (2) Increased tension of the eyeball, sometimes becoming stonelike. (3) Rapid failing of the power of accommodation. (4) Dimness of vision. The pupil is dilated and sluggish. (5) The patient complains of seeing flashes of light and colored halo around a flame or candle. (6) Cupping of the optic disc. (7) Conjunctivitis. The iris also appears steamy. When glaucoma is suspected the patient should be sent to an oculist at once.

The use of atropine causes the iris to crowd into the periphery of the anterior chamber, somewhat occluding the Spaces of Fontana and interfering with the free exit of aqueous humor. As the intraocular tension increases, the stoppage becomes more complete. When the atropine is discontinued the sphincter muscle of the pupil draws the iris away from the Spaces of Fontana and the normal outlet is again opened. As age advances, the sphincter loses its power, and frequently in old people fails to pull the iris away from the Spaces of Fontana, and this condition may result in glaucoma. For the reasons mentioned it is, as a rule, unsafe to use atropine after the ages of from 30 to 35.

Glaucomatous (glau-kom'-at-ous.) Of the nature of glaucoma.

Glioma (gly-oh'-mah). (Gr. glia = glue + oma.) A malignant tumor of the retina.

Gliosarco'ma. Glioma combined with sarcoma.

Globulin (glob'-u-lin). (L. globulus = globule.) A proteid from the lens.

Goggles. Spectacles with wire screens for the eyes.

Goiter (goi'-ter). (L. guttur = throat.) An enlargement of the thyroid gland. **Exophthalmic g.** (See Exophthalmic Goiter.)

Gonorrhe'al Ophthalmia. (Gr. gonos = semen + rhoia = a flow. Ophthalmus = eye.) The most acute form of purulent conjunctivitis. It is caused by the introduction of the urethral discharge to the conjunctival sac.

Graduated Tenotomy. (L. gradus = a degree. Gr. tenon = tendon + tome = incision.) An incomplete cutting of the tendon of an eye muscle.

Granular Lids (Trachoma). (L. granulum, dim of granum = grain.) Roughness and soreness of the inside of the eyelids. This roughness is caused by a swelling of the lymph-corpuscles, forming, as it were, little lymphatic glands or lymphatic follicles.

Gran'ule. A small rounded body. **G. Layer,** one of the layers of the retina.

Gravity (grav'-i-ty). (L. gravitas = heavy.) The state of having weight. (Physics.) The tendency of a mass of matter toward a center of attraction. Specific gravity, the ratio of the weight of a body to the weight of an equal volume of some other body taken as the standard or unit. This standard is usually water for solids and liquids, and air for gases. Thus, 19, the specific gravity of gold, expresses the fact that, bulk for bulk, gold is 19 times as heavy as water.

Gravitation (grav-i-ta'-shun). The act of gravitating. (Physics.) That kind of attraction or force by which all bodies in the universe tend toward each other.

Groove (groov). A furrow, crease or sulcus. A narrow, elongated depression on any surface. **Lachrimal G.,** the bony channel which lodges the lacrimal sac. It is located at the anterior and inner part of the orbit; **cavernous G., carotid sulcus,** the groove on the upper surface of the sphenoid bone, supporting the cavernous sinus and the carotid artery; **optic G.,** a groove on the upper surface of the sphenoid bone between the optic foramen in which rest the optic commissure.

HALLER'S CIRCLES. Arterial and venous circles within the eye.

Halo. (Gr. halos = a circular threshing floor.) A reddish yellow ring surrounding the optic disc.

Ha'lo Glaumato'sus. A whitish ring around the optic disc in glaucoma.

Ha'lo Symptom. Seeing of colored rings around lights. This is a symptom of incipient glaucoma.

Hec'tometer. One hundred meters.

Helcol'ogy. (Gr. helkos = ulcer + logia = study.) Science of ulcers.

Helco'sis. (Gr. helkos = ulcer and suffix osis = condition.) The formation of an ulcer.

Hemeralopia (hem-er-a-lo'-pi-ah). (Gr. hemera = day + alaos = obscure + ops = eye.) Day blindness, better vision in a dim light.

Hemiachromatopsia (hem-i-a-chro-mat-op'-si-ah). (Gr. hemi = half + a = without + chroma color + opsis = vision.) Color blindness in one-half, or in corresponding halves, of visual field.

Hemianopia (hem-i-an-o'-pi-ah), **Hemianopsia.** (Gr. hemi = half + an = without + opsis = vision.) Blindness for one-half the field of vision in one or both eyes.

Hemiopic (hem-e-op'-ik) ("half vision"). That condition of the eye in which only half of the object is seen.

Hemophthal'mia, Hemophthal'mus. (Gr. haima = blood + ophthalmos = eye.) Extravasation of the blood inside of the eye.

Hemorrhage (hem'-or-aj). (Gr. haima = blood + rhagia = to burst.) Escape of blood from the veins or arteries.

Hering's Theory. This is a doctrine which holds that color-perceptions are dependent on a visual substance in the retina, which is variously modified by anabolism for black, green, or blue, and by catabolism for white, red and yellow.

Heterochromia (het-er-o-kro'-me-ah). (Gr. heteros = other + chroma = color.) A difference in color (in the irides or of different parts of the same iris).

Heterometropia (het-er-o-me-tro′-pi-ah). (Gr. heteros = other + metron = measure + ops = eye.) That condition in which the refractive power is unlike in the two eyes.

Heteronymous (het-er-on′-im-us). (Gr. heteronymos = having a different name.) Crossed. See Diplopia.

Heterophoral′gia. Pain with heterophoria.

Heterophoria (het-er-o-pho′-ri-a). (Gr. heteros = other + phoria = tending.) That condition of the eyes in which the visual axes, although parallel when in use for distant vision, deviate in another direction when the extrinsic muscles are in a state of rest. It is subdivided into eight kinds. When the eyes have a tendency to turn in it is known as **esophoria**; if a tendency to turn out, it is known as **exophoria**; if a tendency to turn up, it is known as **hyperphoria**; if up and in, **hyperesophoria**, and if up and out, **hyperexophoria**; if a tendency downward, it is known as **cataphoria**; and if down and in, **esocataphoria**; if down and out, **exocataphoria.** Any error of refraction is liable to bring on **Heterophoria**, and by correcting the error, the **Heterophoria** may disappear, though it may linger for a month or two. Again one muscle may be too short or too long and a prism will have to be worn, thus allowing the eyes to deviate in order to avoid strain.

Heterophthal′mos. (Gr. heteros = other + ophthalmos = eye.) That condition in which the irides differ in color.

Heterotropia (het-er-o-tro′-pi-a). (Gr. heteros = other + trope = I turn.) A condition in which the extrinsic muscles are no longer able to hold the eyes parallel and there is a positive and visible appearance of their deviating. They may turn in any direction, as in **heterophoria**; if upward, **hypertropia**; if downward, **hypotropia** or **catatropia**; if inward, **esotropia**; if outward, **exotropia**. For permanent deviation, see Strabismus.

Hippus (hip′-us). (Gr. hippos = horse so named from its irregular movement.) Spasmodic pupillary movements, independent of the action of light.

Histology (his-tol′-o-je). (Gr. histos = tissue + logia = discourse.) The science of the minute structure and composition of tissues.

Holmgren's Test (holm′-grens). A color test with a number of different colored yarns representing the various shades of different colors. Used for detecting color blindness.

Homocentric Rays (ho-mo-sen′-tric). (Gr. homos = same + kentron = center.) A conic pencil of light rays.

Homonymous. (Gr. homonymos = of the same name.) See Diplopia.

Hordeolum (hawr-dee′-o-lum). (L. hordeum = barley.) Sty; inflammation of sebaceous glands of the eyelid.

Horizon (ho-ri′-zun). (L. horizon = the boundary line.) The circle which bounds that part of the earth's surface visible to a spectator from a given point; the apparent junction of the earth and sky.

Horizontal Line (hor-i-zon′-tal.) A constructive line, either drawn or imagined, which passes through the point of sight, and is the chief line in the projection upon which all verticals are fixed, and upon which all vanishing points are found. **Horizontal plane** is a plane parallel to the horizon, upon which it is assumed that objects are projected.

Horny Epithelium. Trachomatous conjunctivitis.

Horopter (ho-rop′-tur). (Gr. horos = limit + opter = one who sees.) The field of binocular vision as seen with the eyes fixed.

Hot Eye. Temporary congestion of the eye. This is seen in gouty patients.

Humor. (L. humere = to be moist.) A fluid element of the eye. (Aqueous, crystalline lens and vitreous.)

Hutchinson's Pupil. One that is dilated on one side.

Hyaline (hi′-al-in). (Gr. hyalos = glass.) Glassy.

Hyalitis (hy-al-i′-tis). Inflammation of the vitreous humor or hyaloid membrane.

Hyaloid (hy′-al-oid). (Gr. hyalos = glass + eidos = resemblance.) That which resembles glass in its transparent qualities. **Hyaloid membrane** surrounds and encloses the vitreous humor and forms the suspensory ligaments.

Hyaloid Artery. The fetal branch of the central artery of the retina.

Hyaloid Canal, or Canal of Stilling. The canal through the vitreous body, occupied by the hyaloid artery during fetal life.

Hyaloid Fossa. (Gr. hyalos = glass + L. fossa = ditch.) The depression in the anterior surface of the hyaloid membrane in which the crystalline lens lies.

Hyaloid Membrane. The delicate transparent membrane which forms a sac and contains the vitreous humor, and forms the suspensory ligaments of the lens and the Zone of Zinn.

Hydrophthalmia (hy-drof-thal′-mi-ah), **Hydrophthalmus.** (Gr. hydro = water + ophthalmos = eye.) Increase in the fluid contents of the eye.

Hydrops (hi′-drops) (dropsy). An abnormal collection of fluid in any part of the body.

Hygroma (hi-gro′-mah). (Gr. hygros = fluid + oma = tumor.) A sac or cyst filled with fluid.

Hyperaesthesia (hi-per-as-the′-si-ah). (Gr. hyper = overmuch + aisthesis = sensation.) Over-sensitiveness. **H. of Retina,** over-sensitiveness of the retina.

Hyperchromatism (hy′-per-chro′-ma-tism). (Gr. hyper = overmuch + chroma = color.) Having an unusual intensity of color.

Hyperemia (hi-per-e′-me-ah). (Gr. hyper = over + haima = blood.) A condition where there is an abnormal fullness of the blood vessels. **H.** of the eyelids is often a forerunner of inflammation. It is usually accompanied by a slight marginal blepharitis and even conjunctivitis, and if these are relieved the hyperemia to a great extent will disappear.

Hyperkeratosis (hy-per-ker-at-o′-sis). Hypertrophy of the cornea.

Hypermetropia (hy-per-me-tro′-pi-ah). (Gr. hyper = over + metron = measure + ops = eye.) (Far sighted.) An error of refraction, where parallel rays of light focus back of the retina with the muscles of accommodation at rest, due to the shortness of the eye from before, back or insufficient curvature of the dioptric media. Subdivided into three classes—**latent, manifest** and **total. Latent h.** has no subdivisions; it is hypermetropia that is hidden by cramp of the ciliary muscle, and will not relax without the use of drugs at the time of fitting, but when the correction for the manifest is worn, the cramp begins to relax and more hypermetropia becomes manifest. It may take a week or a year. **Manifest h.** is that part found and corrected with the trial case and retinoscope. It is said to have three subdivisions, namely, facultative, relative and absolute. **Facultative h.** is where the patient has the ability to overcome

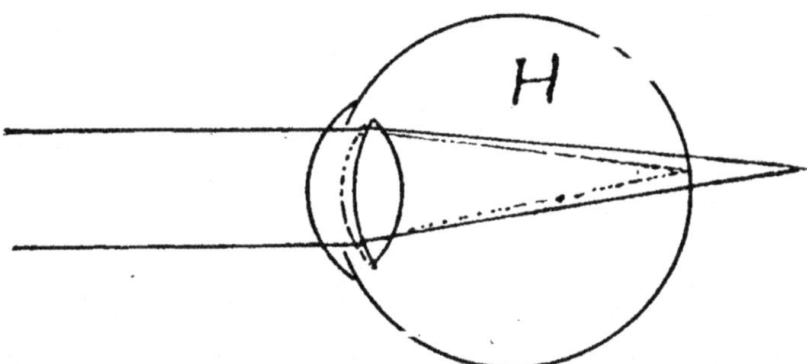

A hypermetropic eye. The heavy lines show the focus of parallel rays behind the retina. The dotted lines show the effect of accommodation upon the same rays.

his error by accommodation, and sees well at all distances. Glasses relieve strain, but do not improve vision in this case. **Relative h.** is where it is possible to accommodate for a near point, by converging to a point still nearer—in fact, by squinting. This eye has blurred vision for close work, and plus spheres improve vision.

Absolute h. is where the error exceeds the amount of the accommodation, and the patient is unable to bring the focus to the retina, and vision is blurred at all distances. The correction always improves vision. **Total h.** is the full amount of hypermetropia the patient has. For instance, we correct the eye with the trial case and find 2-D. of manifest; then by the use of drugs relax any cramp; and now find that the same eye has 6-D. of hypermetropia. 4-D. was hidden by cramp. This we call latent, 6-D., being the total amount of hypermetropia.

Hyperope (hi'-per-op). A person who has hypermetropia.

Hyperopia (hi-per-o'-pe-ah). (Gr. hyper = over + ops = eye.) See **Hypermetropia.**

Hyperphoria (hy-per-fo'-ri-ah). (Gr. hyper — above + phoria = tending.) That condition in which one of the eyes, although parallel with its fellow when in use for distant vision, deviates upward when the extrinsic muscles are in a state of rest.

Hyperplasia (hi-per-pla'-ze-ah). (Gr. hyper = above + plasis = a moulding). Excessive tissue formation.

Hypertrophy (hy-per'-tro-fy). (Gr. hyper = above + trophe = nourishment.) An abnormal increase in the size of a part or an organ.

Hypertropia (hy-per-tro'-pi-ah). (Gr. hyper = above + trope = turn.) Elevation of one visual axis above the other.

Hyphemia (hi-fe'-me-ah). (Gr. hypo = below + haima = blood.) **Hemorrhage** within the eye.

Hypnogenet'ic. (Gr. hypnos = sleep + genesis = production.) Causing or producing sleep.

Hypnolepsy (hip'-no-lep-se). (Gr. hypnos = sleep + lepsis = a seizing.) Abnormal sleepiness.

Hypometropia (hy-po-me-tro'-pi-ah). (Gr. hypo = under + metron = measure + ops = eye.) See Myopia and Brachymetropia.

Hypophoria (hi-po-fo'-re-ah). (Gr. hypo = below + phoria = tending.) That condition in which one of the eyes, although parallel with its fellow when in use for distant vision, deviates downward when the extrinsic muscles are in a state of rest.

Hypopyon (hi-po'-pe-on). (Gr. hypo = beneath + pyon = pus.) Pus in the anterior chamber of the eye.

Hypotenuse (hi-pot'-e-nus). (Gr. hypo = under + teinein = to stretch.) The side of a right-angled triangle opposite the right angle.

Hypotonia (hi-po-to'-ne-ah). (Gr. hypc = under + tonos = tone.) Diminished intraocular tension.

Hypotonus (hi-pot'-o-nus). See **H**ypotonia.

Hypotony (hi-pot'-o-ne). See Hypotonia.

IDENTICAL POINTS. When the image falls on corresponding points on the retinae of the two eyes.

Illaqueation (il-lak-we-a'-shun). (L. illaqueare = to ensnare.) The curing of ingrowing eyelashes by drawing with a loop.

Illumination (il-lu-min-a'-shun). (L. illuminare = to light up.) The lighting up of a place or object for inspection. **Focal i.**, when light is brought to a focal point by lens or mirror. **Axial i.**, when light is transmitted or reflected along the axis of a lens. **Direct i.**, light thrown directly upon the object. **Oblique i.**, when an object is illuminated from one side.

Illusion (il-lu′-shun). (L. illudere = to mock.) An unreal image presented to the mental vision.

Image (im′-ej). (L. imago = likeness.) A picture or conception of anything real. **Aerial i.**, image seen as in the air by the ophthalmoscope. **Direct i.**, **Erect i.**, and **Virtual i.**, formed by rays not yet focused. An upright image. **False i.**, image formed on the retina of the deviating eye in strabismus. **Optical i.**, an appearance of an object created by refraction or reflection.

Imbalance. That condition in which the eyes tend to deviate from parallelism with the extrinsic muscles in a state of rest. See heterophoria

Inad′equacy. (L. in = not + adaequare = to be equal.) Unable to perform allotted function.

Incident. (L. incidere = to fall into or upon.) Falling or striking upon, as a ray of light upon a reflection surface.

In′cident Ray. The name given to a ray of light before it strikes the second medium.

Index of Refraction. The refracting or bending power of the medium as compared with air, the normal standard, and the index of which is the unit 1. Water as compared with air is 1.33; crown glass, 1.52; flint glass, 1.62 + ; pebble, 1.54; diamond, 2.4, the greatest index of any known medium. The transparent parts of the eye in their order are as follows: the cornea, 1.33; the aqueous humor, 1.33; the crystalline lens, 1.43, and vitreous humor, 1.33. Different indices of refraction would mean different densities.

Induction (in-duk′-shun). (Physics.) The property by which one body, having electrical or magnetic force, induces it in another body without direct contact.

Inertia (in-er′-shi-a). (L. idleness.) (Physics.) That property of matter by which it tends when at rest to remain so, and when in motion to continue in motion, and in the same straight line or direction unless acted upon by some external force.

Infiltration (in-fil-tra′-shun). (L. in, and filtrare = to filter.) The act or process of infiltrating a fluid into the cellular tissue.

Infinite Distance. When rays of light proceed from a distance of twenty feet or more they are considered parallel. and are said to come from infinity.

Inflammation (in-flam-ma′-shun). (L. inflammare = to burn.) A diseased condition characterized by redness, pain, heat and swelling. **Traumatic i.**, that which follows a wound or injury.

Inflection (in-flek′-shun). (L. in = in + flectere = to bend.) The act of bending inward or that state of being bent inward.

Infra. A prefix denoting a position below the part denoted by the word to which it is joined.

Infraduction, Deorsumvergence. The act or power of turning one eye downward from its fellow.

Infraorbital (in-fra-or′-bi-tal). Situated beneath the orbit.

Innervation (in-nerv-a′-shun). (L. in = not + nervus = nerve.) The sending of nervous stimulus or power to an organ through its nerves.

Innervate (in-nerv′-et). To supply with nerves; to give nervous stimulus to.

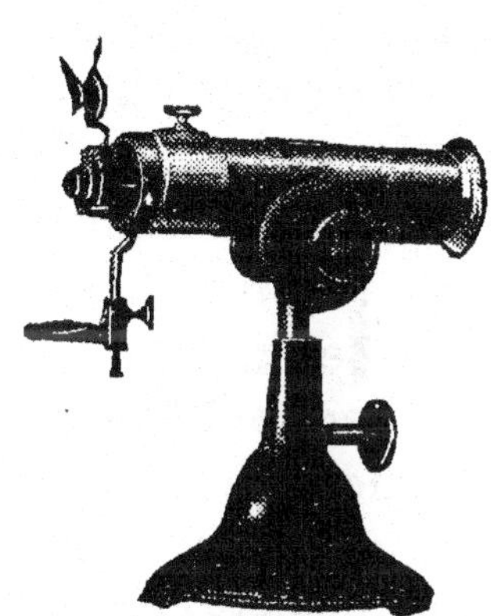

INSTRUMENTS AND THEIR USES

Amblyoscope. An instrument to stimulate, exercise and develop the fusion faculty in strabismus, or squinting patients.

Color Test (Holmgren's). A set of worsteds, consisting of various shades and tints, for testing color blindness.

Focimeter (fo-sim′-e-ter). An instrument for measuring the focal lengths of lenses or combination of lenses.

Keratometer. See Ophthalmometer.

Kryptoscope. An instrument used for testing strain in ophthalmic lenses. With this instrument strain can be detected in fused (kryptok) lenses or when the screw holding a lens in its frame is too tight.

Latest Optometer. An instrument combining the advantages of a fixed and revolving cell trial frame, Stevens Phorometer, Rotary Prism and Maddox Multiple Rod.

Myometer (my′-o-meter). An instrument for diagnosing and correcting muscular insufficiency at the near point of vision.

Ophthalmoscope. An instrument with which the interior of the eye may be examined. Also the dioptric and pathological states may be determined. There are many different kinds of ophthalmoscopes; for instance, the Loring is a small hand affair, which contains a mirror and a number of lenses; the self-luminous, by DeZeng, also a hand instrument; and the combined ophthalmoscope and retinoscope, a combined instrument for indirect ophthalmoscopy and for retinoscopy. This is a large machine which stands on a table.

Ophthalmometer or Keratometer. An instrument for determining the amount and axis of corneal astigmatism, an objective test.

Ophthalmometroscope. An ophthalmoscope with an attachment for measuring the refraction of the eye.

Perimeter. An instrument for measuring the visual field.

Punctumeter. A simple instrument for determining the far point and the near point. therefore the amount of hypermetropia, myopia, or presbyopia. It also indicates the age of the patient.

Savage Monocular Phorometer and Cyclo Phorometer. Two instruments which together make a complete appliance for measuring all of the muscles of the eye.

Skiascope. A frame with a series of plus and

minus spherical lenses, to be used in place of test frame and lenses when refracting a patient by retinoscopy.

Stevens Phorometer. An instrument for measuring muscular imbalance.

Stigmatometer. An instrument for testing refraction of the eye by the objective method. Also a complete ophthalmoscope for the direct examination.

Insufficiency. Incapacity of normal action within the eye.

Integer (in′-te-ger). (L. a whole number.)

Intercilium (in-ter-sil′-e-um). (L. inter = between + cilium = eyelid.) The space between the eyebrows.

Interorbital (in-ter-or′-bi-tal). Situated between the orbits.

Inter′nus. Internal. The internal rectus muscles of the eye.

Interval. (L. inter = between + vallum = wall.) **Sturm's,** or Focal i. In astigmatism, is the distance between the two foci, at which the principal meridians meet.

Intraocular (in-trah-oc′-u-lar). Situated within the globe of the eye.

Intraocular Tension. Pressure from the fluids within the eye.

Intraorbital (in-trah-or′-bit-al). Situated within the orbit.

Involution (in-vo-lu′-shun). (L. involvere = to roll up.) Multiplication of a quantity into itself any number of times.

Ir′idal. Pertaining to the iris.

Iridectome (ir-id-ek'-tom). An instrument used in cutting the iris in iridectomy.

Iridectomize (ir-id-ek'-tom-ize). To cut away a part of the iris.

Iridectomy (ir-id-ek'-to-my). (Gr. iris + ektome = excision.) The operation for removing a piece from the iris for the relief of tension of the eyeball in the case of glaucoma, thus producing an artificial pupil.

Iridencleisis (ir-id-en-cli'-sis). (Gr. iris + enkleio = to enclose.) An operation for displacing the pupil from its natural position, brought about by drawing the iris into a wound made near the periphery of the cornea, and causing it to become adherent there.

Irideremia (ir-id-er-e'-mi-ah). (Gr. iris + eremia = to deprive.) Defect or imperfect condition of the iris.

Irides (ir'-id-ez). Plural of iris.

Iridesis (ir-id'-e-sis). (Gr. iris + desis = to bind.) Strangulation of a part of the iris to form an artificial pupil.

Iridescent Vision. (Gr. iris = rainbow.) That condition in which variously hued borders are seen surrounding artificial light.

Iridic (i-rid'-ik). Pertaining to the iris.

Iridoavulsion (ir'-i-doh-a-vul'-shun). A term applied to the total removal of the iris when it is completely torn from its periphery.

Iridocele (i-rid'-o-sele). (Gr. iris + kele = hernia.) Hernial protrusion of a slip of the iris.

Iridochoroiditis (ir-id-o-ko-roid-i'-tis). Inflammation of the iris and choroid.

Iridocinesis (ir-id-o-sin-e′-sis). The movement of the iris in contracting and expanding.

Iridocyclitis (ir-id-o-syc-li′-tis). (Gr. iris + kyklos = circle + itis = inflammation.) Inflammation of the iris and ciliary body.

Iridod′esis. That condition in which a loop of iris is drawn out, and strangulated by a fine ligature tied around it over the incision; the little loop soon drops off, and the result is a pear-shaped pupil, with its broad end toward the center.

Iridodialysis (ir-id-o-di-al′-ys-is). (Gr. iris + dialysis = separation.) Separation of the iris from the ciliary body.

Iridodonesis (ir-id-o-do-ne′-sis). (Gr. iris + doneo = agitation.) Trembling condition of the iris.

Iridoncus (ir-id-on′-kus). (Gr. iris + onkos — swelling.) A tumor or swelling of the iris.

Iridoperiphacitis (ir-id-o-per′-i-fa-si′-tis). (Gr. iris + peri = around + phakos = lens.) Inflammation of the capsule of the lens of the eye.

Iridoplania (ir-id-o-pla′-ni-ah). (Gr. wandering.) Trembling of the iris; iridodonesis.

Iridoplegia (ir-id-o-ple′-gi-ah). (Gr. iris + plege = stroke.) Paralysis of the iris. Without defect of accommodation, it usually affects only the action to light, reflex iridoplegia, the associated action remaining. It occurs as a very early symptom in locomotor ataxia, sometimes without any other symptoms of that disease, and should always lead to full investigation. It is probably due to degeneration in that part of the nucleus of the third nerve which presides over the reflex action of the pupil.

Iridorrhexis (ir-id-or-rhex'-is). (Gr. iris + rhexis = rupture.) Rupture of the iris. Tearing away of the margin of the iris.

Iridosclerot'omy. (Gr. iris + tome = incision.) Puncture of the sclerotic and of the edge of the iris.

Iridotomy (ir-id-ot'-o-my) (Gr. iris + tome = incision.) The operation whereby an artificial pupil is formed by the natural gaping of a simple incision in the iris. Iridotomy is most useful when the iris has become tightly drawn toward the operation scar by iritis occurring after a cataract has been removed.

Iris. (Gr. rainbow.) So called from its resembling the rainbow in its many colors. The membrane, stretched vertically in the anterior part of the eye, in the aqueous humor, in which it forms a flat circular partition separating the anterior from the posterior chamber. It is the anterior part of the second tunic, and is perforated by a circular opening called the pupil, which is constantly varying in size, owing to the contractions of its two sets of muscles. Its posterior surface is covered with a black coat of pigment which continues backward over the ciliary body and choroid. The greater circumference of the iris is adherent to the ciliary body and to the sclerotic by the ciliary ligament. (Ligamentum Pectinatum Irides.) Its arteries are from the long ciliary arteries, which form two circles, one broad near its circumference, the other small and seated around the circumference of the pupil. Its veins empty into the long ciliary veins and into the Vena Vortisosea. The pupil is contracted by the circular or sphincter muscle supplied by the motor oculi (3d) nerve and dilated by the radiating muscle or dilator, which is chiefly supplied by the sympathetic. The iris gives the eye its color, regulates the amount of light which enters and prevents spherical aberration of the crystalline lens.

Iris Shadow. The test for maturity, or ripened cataract; created by oblique illumination.

Iritic (i-rit′-ik). Pertaining to the iris.

Iritic Angle. The angle formed by the junction of the iris and cornea.

Iritis (i-ri′-tis). (Gr. iris + itis = inflammation.) Inflammation of the iris, which is usually caused by certain specific blood diseases. It often occurs in the course of ulcers and of wounds and other injuries of the cornea; also with sclerotitis and keratitis.

Irregular Astigmatism. See Astigmatism.

Irritant. (L. irritare = to provoke.) Causing irritation.

Ischemia (is-ke′-me-ah). (Gr. ischo = restrain + haima = blood.) Bloodlessness.

Ischemia Retinae (is-ke′-me-ah). Diminution of arteries in the retina.

Iso. (Gr. isos equal.) A prefix denoting equality.

Isocoria (i-so-co′-ri-ah). (Gr. isos = equal + kore = pupil.) Where the pupils in the two eyes are equal.

Isometropia (i-so-met-ro′-pi-ah). (Gr. isos = equal + metron = measure + ops = eye.) The state in which both eyes are alike in their refraction.

Isosceles (i-sos′-e-lez). (Gr. isos = equal + skelos = leg.) Having two sides equal.

Isotropic (trop′-ik). (Gr. isos =equal + trope = a turning.) Equal in refractive power.

JAGER'S TEST TYPE. The standard type for close reading, a hand chart.

Joffroy's Symptom. That condition which exists when patient suddenly turns his eyes upward and there is absence of facial contraction; seen in exophthalmic goiter.

KATAPHORIA (kat-a-phor′-ia). (Gr. kata = down + phoria = tending.) That condition in which the eyes turn downward when the extrinsic muscles are in a state of rest. Stevens gives 50° for the maximum depression of normal eyes

Keratalgia (ker-at-al′-je-ah). (Gr. keras = horn + algos = pain.) That condition in which there is pain in the cornea.

Keratectasia (ker-at-ek-ta′-si-ah). (Gr. keras = horn + ektasis = extrusion.) That condition in which the cornea protrudes.

Keratitis (ker-at-i′-tis). (Gr. keras = horn + itis = inflammation.) Inflammation of the cornea.

Keratocele (ker-at′-o-cele). (Gr. keras = horn + kele = hernia.) Corneal protrusion of Descemet's Membrane.

Keratoconus (ker-at-o-ko′-nus). (Gr. keras = horn

\+ konos = cone.) That condition in which there is a conical cornea.

Keratoglobus (ker-at-o-glo′-bus). (Gr. keras = horn + L. globus = ball.) A globular protrusion of the cornea.

Keratohelcosis (ker-at-o-hel-ko′-sis). (Gr. keras = horn + helkosis = ulceration.) That condition wherein there is ulceration of the cornea.

Keratoiri′tis. (Gr. keras = horn + itis = inflammation.) That condition wherein the cornea and iris are inflamed.

Keratoma. (Gr. keras = horn + oma = tumor.) A horn-like tumor or swelling.

Keratomalacia (ker-at-o-ma-la′-she-ah). (Gr. keras = horn + malakia = softness.) Softening of the cornea.

Keratome (ker′-at-om). A knife for incising the cornea.

Keratometer (ker-at-om′-e-ter). (Gr. keras = horn + metron = measure.) An instrument used for measuring the cornea. It is commonly called the ophthalmometer, of which there are several different makes.

Keratometry (ker-at-om′-e-try). Measurement of corneal curves.

Keratomycosis (ker-at-o-my-ko′-sis). (Gr. keras = horn (cornea) + mykes = fungus.) Fungous disease of the cornea.

Keratonyxis (ker-at-o-nik′-sis). Gr. keras = horn + nyxis = a pricking.) Puncture of the cornea.

Keratoplasty (ker′-at-o-plas-ty). (Gr. keras = horn + plasso = I form.) Plastic surgery of the cornea.

Keratoscleritis (ker-a-to-skle-ri′-tis). (Gr. keras = horn + sclera + itis.) Inflammation of both cornea and sclera.

Keratoscope (ker′-at-o-scope). (Gr. keras = horn + skopeo = I examine.) Instrument for examining the cornea.

Keratoscopy (ker-at-os′-ko-pe). Examination of the cornea with a keratoscope. Skiascopy.

Kerectomy (ke-rek′-to-me). (Gr. keras = horn + ectome = excision.) Removal of part of the cornea.

Kilometer. One thousand meters.

Kopiopia or **Copiopia** (ko-pee-oh′-pee-ah). (Gr. kopos = fatigue + ops = eye.) See Asthenopia.

Korectomia or **Corectomia** (ko-rek-to′-mee-ah). (Gr. kore = pupil + extome = excision.) The operation for artificial pupil by removal of a part of the iris.

Korectopia (kor-ek-to′-pe-ah). (Gr. kore = pupil + ektopos = out of place.) Displacement of the pupil.

Koroscopy (ko-ros′-ko-pee). (Gr. kore = pupil + skopeo = I view.) See Retinoscopy.

Kryptok. (L. crypta = a vault; a hidden place.) The name applied to a bifocal lens made by fusing two pieces of glass of different density together, so as to become one integral piece with no visible line of demarcation.

LACHRYMAL (lak′-rim-al). (L. lacrima = a tear.) Pertaining to tears

Lachrymal Apparatus. Consists of the lachrymal

gland which secretes the tears and the exsecretory ducts which convey the fluid to the surface of the eye. This fluid after passing over the eye

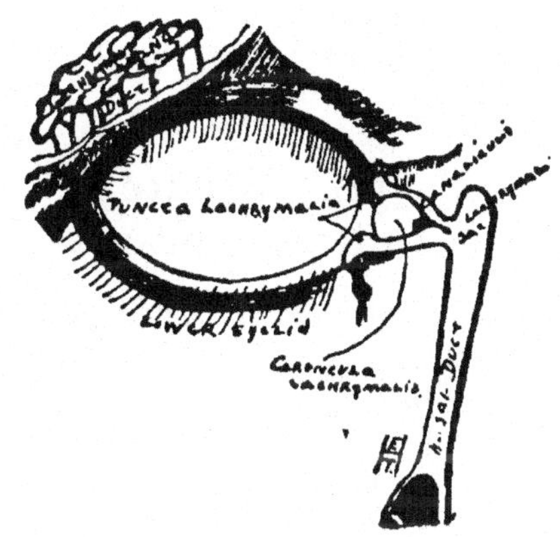

runs through the puncta into the lachrymal canal, then to the lachrymal sac and along the nasal duct into the cavity of the nose.

Lachrymation (lak-rim-a′-shun). The secretion and discharge of tears.

Lachrymotomy (lak-rim-ot′-o-my). (L. lacrima a tear + G. tome = incision.) Operation for incision of lacrimal duct or sac.

Lacrimal. Gland. See Gland.

Lacrimal, Lacrymal. Same as Lachrymal.

Lacu′nar Orbitae. The roof of the orbit of the eye.

Lacus. (L. "a lake.") The small circular portion at the nasal side of the opening between the eyelids.

La′cus Lacrima′lis. (L. "lake" + lacrima = a tear.) The triangular space at the inner canthus between the two eyelids.

Laevophoria (le′-vo-phor′-ia). (L. laevus = left + G. phoria = tending.) That condition in which the eyes turn to the left when the extrinsic muscles are in a state of rest.

Lagophthalmus (lag-of-thal′-mus). That condition in which it is impossible to close the eyes.

Lamella (lam-el′-ah). (L. dim. of lamina = plate.) A thin plate or scale. Lamina.

Lamina (lam′-in-a). (L. "a plate.") A layer consisting of a flat, thin membrane.

Lamina Cribrosa (lam′-in-a crib-ro′-sa). (L. "a plate" + cribrum = sieve-like.) The perforated area in the sclerotic of the eye through which the optic nerve fibers pass to form the retina.

Lamina Fus′ca. (L. "a plate" + fuscus = brown.) The outside layer of the choroid.

Landolt, Edmund, M. D. Ophthalmologist, born in Aaran, Switzerland, in 1846; pursued his professional studies in the universities of Heidelberg, Vienna, Berlin, Utrecht and Zurich, graduating from the latter in 1869; then worked for more than a year as Horner's assistant in the Zurich clinic for eye diseases; in 1874 he established himself in Paris as an ophthalmologist. His investigations in his specialty have been distinguished by their originality. Among his works are "On the Retina," "A Manual of Ophthalmoscopy," published in French, English, German and Spanish; "The Refraction and Accommodation of the Eye."

Lapsus (lap′-sus). The dropping of the upper lid, produced by a paralysis of the levator palpebra muscle. Synonym, Ptosis.

Lashes. The name given to the hairs of the eyelids.

Latent (la'-tent). (L. latere = to be concealed.) That which is not apparent or manifest. See Hypermetropia.

Layer. A stratum having a certain amount of thickness and serving the purpose of a covering.

Leber's Disease (La'-berz). (Theodor Leber, German Ophthalmologist, 1840.) Atrophy of the optic nerve, which is hereditary.

Lema (le'-ma). The dry, hard, yellowish incrustations which collect in the inner canthus.

Lens. (L. "a lentil.") The term lens was first applied to any transparent refracting body which had two spherical surfaces, on account of its resemblance to a vegetable known as a lentil. A lens is a transparent substance, crown or flint glass chiefly, ground with regular curvature on one or both of its opposite sides, but not parallel to each other, through which an object may appear to be increased or decreased in size, and may have either convex or concave spherical or cylindrical surfaces. There are six varieties of spherical lenses—three plus and three minus—all of which can be made the same dioptric power, the only difference being in the shape of the lens. Plus or positive lenses are thickest in the center. Minus or negative lenses are thinnest at their centers. A plus sphere will refract the same in all its meridians and converge parallel rays of light to a point of focus. while a minus sphere will diverge parallel rays from a point. The different forms of plus and minus spherical lenses are here represented:

A line passing through the optical center at right angles to the surfaces of these lenses is not refracted, and is known as the principal axis, while all other rays undergo more or less refraction. A secondary axis is any line which

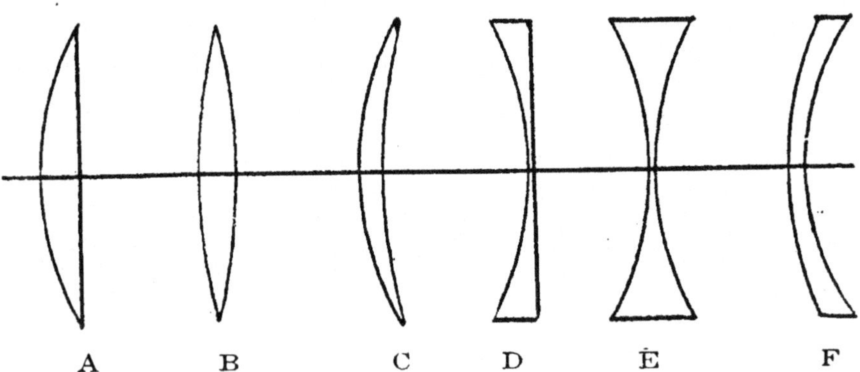

A Plano Convex.
B. Biconvex.
C. Periscopic Convex.
D. Plano Concave
E. Biconcave.
F. Periscopic Concave.

crosses the principal axis at the optical center of a lens. It is not a straight line, but a refracted one, and on emerging takes a direction parallel to that which it would have pursued had it not been interrupted by the lens. **Achromatic L.**, a lens composed of two pieces, one of crown and the other of flint glass; the former one being plus and the latter minus, and only half as strong in its refractive power, but of equal dispersive power, and overcomes chromatic aberration. **Aplanatic L.** is on the order of the achromatic lens, except that the minus is divided and placed half on each side of the plus. In this way not only the chromatic but the spherical aberration is overcome, and a perfect lens formed. They are used for high-power instruments. **Bifocal L.** (see Bifocal.) **Composite L.**, a lens having three features, namely, spherical, cylindrical, and prismatic. **Cylindrical L.**, a lens with refractive power in all meridians but one. This one is known as the axis, and is nothing more than plano glass. The refraction varies from zero at the axis to the full strength, which is at right angles to the axis. **Crossed L.,** a double-convex lens with one radius equal to six times the other. **Crystalline L.,** the lens of the eye which resembles a crystal. A transparent double-convex lens situated in its capsule behind the pupil, between the aqueous and vitreous humor, and when in a state of rest has a focal strength of from plus 19 to plus 20 diopTries. **Compound L.,** a lens consisting of two or more lenses made up together, such as a sphere and a cylinder. **Toric L.,** a lens with power in all meridians, but of different amounts on the same side, usually made extra deep periscopic. **Fresnel L.,** a compound lens formed by placing around a central convex lens, rings of glass so curved as to have the same focus; used, especially in light houses, for concentrating light in a particular direction. **Lenticular L.** is a lens which is plano at the edges, and the power is ground in a space of about half an inch in diameter in the center. When a plus lens is required it is made in the form of a scale and cemented on a plano or simple cylinder. In this way we do away with the thick edge of a high-power minus lens, and it also makes up in a thinner form for a high-power plus, but they are never made up in weak lenses. **Orthoscopic L.,** a lens with two elements, a sphere and a prism, so arranged that the amount of accommodation and convergence used should exactly correspond. **Periscopic L.,** a lens having a convex and concave surface. **Ret'roscopic L.,** a lens that is tilted inward at the top.

(OVER)

Toric Lenses. The word toric was taken from the word torus, which means in architecture the large semicircular molding used in the bases of columns, and the term is applied to a lens having curvature in all meridians, but of different amounts, on the same side of the lens with its meridians of greatest and least curvature at right angles to each other. The meridian of least curvature is known as the base curve, while the other side of the lens will be plano, a concave or a convex sphere; but usually made concave.

To give an idea of the appearance and proper uses of such lenses, I will put up a prescription for one of the five subdivisions of ametropia which can be corrected by lenses in toric form. This can best be explained by diagrams.

This prescription will be for compound hyperopic astigmatism + 3. sph. ◯ + 2. cyl. ax. 90.

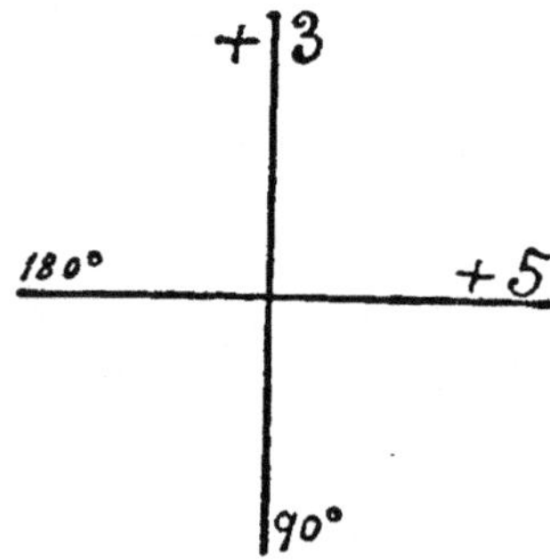

This lens when ground must refract plus three dioptries in the ninetieth and plus five dioptries in the one hundred and eightieth meridians, independent of its shape.

But in order to get a deep periscopic effect, the advantage of which I will explain later, suppose we grind one side of the lens thus:

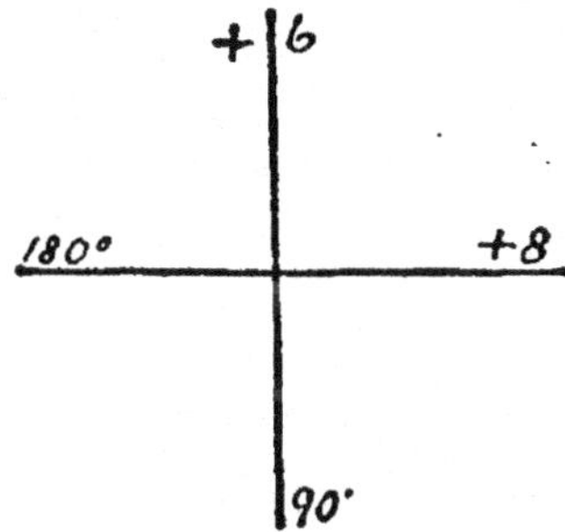

The difference between the curvatures in the two meridians gives us the desired value of the cylinder, and on the other side we will grind a minus three sphere, which will neutralize plus three from all meridians, leaving the lens with the required strength. In this lens you get a plus sphere and a plus cylinder, and at the same time, if a cement scale is required, it can be placed next to the eye.

When a toric lens is desired, it is not necessary for the refractionist to mention the curvature. For the sake of simplicity, just write the word "toric" beneath the description of the lenses in your prescription. Then write the prescription in the usual way.

Toric lenses are more expensive than the old form of lenses, but on account of their superiority they are coming more into general use.

In the first place, they allow a greatly enlarged field of vision, by allowing the patient to roll the eye and at the same time see through the edges of his lens.

With lenses of the ordinary type, when an eye turns it looks obliquely through them and obtains a prismatic effect that is not desired, causing the image to be more or less distorted on the retina, and at the same time the patient is bothered with reflection from the back of the lenses of objects on the side. With deep periscopic lenses the curve coincides approximately with the arc formed by the eye in turning, and

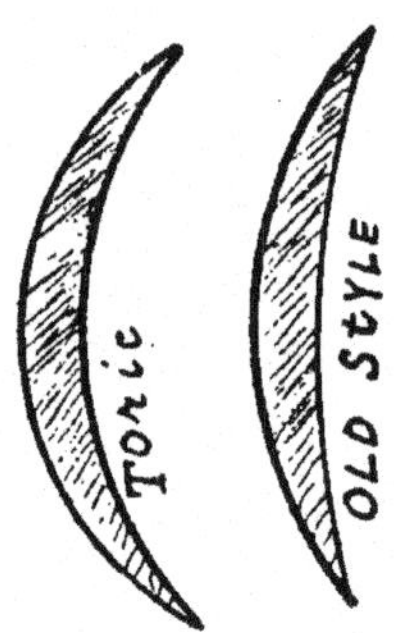

the eye is looking much more directly through the lens and obtains a much larger field of vision without the extra prismatic effect.

The diagrams show the shape of the two kinds of lenses from the same prescription, + 2 sph. ⊂ — 1 cyl. ax. 90.

Again, the edges of the lenses come nearer to the face, thus adding to the patient's appearance. The result of a lens of this description is freedom and comfort to the wearer, so much so that the extra cost should not be considered.

When a cement bifocal is required, the toric side should always be plus, so that a minus sphere will be next to the eye, on which the scale may be cemented.

Lens Capsule (L. a lentile + dim. of capsa = box). A transparent, highly elastic, and brittle membrane which encloses the crystalline lens. It rests in a depression of the vitreous body just behind the iris, and is held in position by the suspensory ligaments.

Lenticonus (len-tik-o'-nus). (L. lens + conus =

cone.) Exaggerated curvature of the crystalline lens.

Lenticular (L. lenticula = a lentile). Resembling a lentil. Lens-shaped; pertaining to a lens.

Lenticular Astigmatism. See Astigmatism.

Lesion (le′-shun). (L. loedere = to injure.) Any hurt, wound, or local degeneration.

Leucoma (lew-ko′-mah). See Leukoma.

Leukoma (lew-ko′-ma). (Gr. leukos = white.) White corneal opacity. Albugo.

Levator (L. levare = to lift). Elevator; a muscle raising a part.

Levator Palpebra Muscle (le-va′-tor pal′-pe-bra mus′-l). See Muscles.

Ligament (lig′-a-ment). (L. ligare = to bind.) A tough band of connective tissue, the purpose of which is to connect the bones together or surround them as a capsule. There are several ligaments concerned in the anatomy of the eye. Check Ligament, Ciliary Ligament, Palpebral Ligament, External Palpebral Ligament, Internal Palpebral Ligament, Lockwood Ligament, Suspensory Ligament or Zone of Zinn, Ligament of Zinn. The **Ciliary L.**, or circle (annulus albidus), is the bond of union between the external and middle tunics of the eyeball, and serves to connect the cornea and sclerotic, at their line of junction, with the iris and external layer of the choroid. It is also the point to which the ciliary nerves and vessels proceed previously to their distribution, and it receives the anterior ciliary arteries through the anterior margin of the sclerotic. A minute vascular

canal is situated within the ciliary ligament, called the ciliary canal, or the Canal of Fontana, from its discoverer. **Check L.**, the fibrous bands attached by one end to the anterior wall of the orbit and by the other to the tendons of the recti muscles. Those on the inner side are called the internal check ligaments, and those on the outer or temple side the external check ligaments. The action of these ligaments is a normal one. They prevent or retard overaction of the abductors or adductors. **Lockwood L.**, a band of orbital tissue in the anterior part of the orbit, where it forms a hammock-like band attached at one end to the lachrimal and at the other to the malar bone. Its broader central part passes below the eyeball over the Capsule of Tenon. The **Palpebral L.** joins the cartilage of the lids to the orbit the same as the tarsal ligament. The **External Palpebral L.** unites the lid to the outer edges of the orbit. The **Internal Palpebral L.** covers an area including the upper maxilla to the inner margin of the lid. The **Suspensory L.**, or Zone of Zinn, surrounds the crystalline lens and holds it in place within the circle of the muscle of accommodation.

Ligament of Zinn. A circular ligament which is attached to the bone at the optic foramen, from which arise the four recti muscles.

Ligamentum Pectinatum. The ligaments which pass from the base of the iris to the cornea. Through its meshes pass Fontana's spaces.

Light (L. lux = light). Light is that physical force which, after entering the eye and acting upon the sensitive elements of the retina, excites in the mind the impression of vision (or vibrations of ether). It is an extremely rare, elastic medium which is diffused over the universe, emanating from the sun and stars, bodies in a state of ignition, candle flame, electricity, etc. It is said to travel at the velocity of 186,000 miles per second while in air. Its speed is retarded when it enters a denser medium, as water, glass, etc. The amount of the retardation depends on the density of the medium. It regains its former speed on emerging into air again. The unit of light is 1 candlepower.

Bodies sending forth rays or waves of light are called luminous; and those through which it passes easily, transparent; those through which it can pass less easily, translucent; and those through which it cannot pass, opaque. When light meets the surface of a body it may be reflected, absorbed, refracted, or diffracted.

It is the cause of color of all bodies, being entirely reflected by white objects and absorbed by black. It is decomposed in passing through a prism, and its seven primary colors exposed, thus: violet, indigo, blue, green, yellow, orange, and red. Of these, violet is refracted the most and red the least.

Rays of light that do not enter the eye are invisible. A ray entering a darkened room is visible because the floating dust reflects some of it to the eye.

White light is the combination of all colors,—color is not a property of matter, but of light,—for instance, the color of a flower is its property of reflecting to the eye light of that particular color, the other colored rays being absorbed. This is what is known as **selective absorption.**

Different colored light has different wave length. Red has the longest and violet the shortest waves. It will be noted that violet must travel much faster than red in order to travel with it through space. The main primary colors are red, yellow and blue; the others, orange, green, indigo and violet, are secondary or "binary" colors.

In moving colored objects so that their image moves from the macula to the periphery of the retina, colors grow dim; green is lost first, then red, yellow and blue last. Color blindness is generally congenital, but may be acquired by changes in the retina, particularly in atrophy of the optic nerve.

From an optical standpoint there are two theories of the way in which light travels, namely, rays and waves.

A **Ray** is the smallest visible line of light.

A **Beam** is a collection or bundle of parallel rays.

A **Pencil** is a number of converging or diverging rays.

Rays emanating from an illuminating or an illuminated point always diverge; in nature there are no converging rays, neither are there any absolutely parallel, but those proceeding from a point 20 feet or farther away are so nearly so that the difference can only be mathematically expressed, and for the purposes of optics are considered as parallel. According to the calculations of astronomers, light moves at the rate of about 186,000 miles in a second; according to this, it requires about nine minutes for the waves of light from the sun to reach the earth, and those from the nearest fixed star are five years on their journey before they reach us.

From an optical standpoint we now refer to the "Wave Theory," and in order to do this it will be necessary to draw somewhat on one's imagination. You have ofttimes noticed when a stone is dropped into a calm pond of water it throws forth circular waves in all directions. The first or nearest wave to the stone will have the shortest radius of curvature, or, in other words, the greatest strength of curvature. As this wave spreads it will decrease in curvature until it has traveled 20 feet. Beyond 20 feet the waves are considered plane, meaning by this that, when on account of the pupil of the eye being about an eighth of an inch in diameter, we cut from a wave of light that has traveled 20 feet a piece one-eighth of an inch long, that is, the amount that would enter the eye, it would have so slight a curve that it is considered to have none. This is known as a plane wave.

The word minus denotes less; the farther the wave travels from its center, the less its curvature; therefore all waves that are going from a point we consider minus, and for the sake of simplicity we must compare the waves of light with the waves of water, and instead of dropping the stone we will light a candle that will throw off waves in all directions. When a wave has traveled one-half inch from a point it has a curve of minus 80, because it has a radius of curvature of one-eightieth meter. Now, as the one and same wave moves on, it loses its curvature; thus, when it has traveled one inch from its center its curvature is less, or minus 40; and at two inches, minus 20; three inches, minus 13; four inches, minus 10; five inches, minus 8; twenty inches, minus 2; forty inches, minus 1; eighty inches, minus .50 (these figures are the fractional part of a meter, which the distance represents); twenty feet, no curve, or plane wave. Now, if one will stop to think, he will observe these figures compare with the focal length of lenses in the trial case; that is to say, a wave that has traveled forty inches from a point is known as a minus 1, and a 1-dioptry lens has a focal length of forty inches. A wave that has traveled twenty inches from a point is known as a minus 2, while a 2-dioptry lens would focus at twenty inches. For instance, you may ask yourself, "What would be the curvature of a wave of light that has a radius of thirteen inches?" You would at once think of the dioptric number of the lens that would focus at thirteen inches. This would be a 3-dioptry. Then you would say that the curve is minus 3, if it is going from a point, but if going to a point, plus 3. You will notice that in referring to a meter it is spoken of as forty inches. There is a difference between the two, yet it is near enough for our purpose, and saves the trouble and inconvenience of working with fractions; so far, we have spoken of the minus wave, as all waves in nature are minus; in order to have a plus wave we must use artificial means, and will work out the following example: Place a lighted candle forty inches from a plus 3 sphere; considering the candle the point from which the light comes; the wave has traveled from a point forty inches before it enters the lens, therefore it enters a minus 1 wave. Minus and plus neutralize. If more plus than minus is present there will remain, after neutralization, an amount of plus equivalent to the difference. Therefore the minus 1 will go through the plus 3 sphere, and will emerge a plus 2 wave, and focus at twenty inches; at the focus they will cross and begin to diverge, or rather become minus.

Light travels at the rate of 186,000 miles in a second while in air, but in passing through a denser medium, such as glass, its speed is retarded according to the density, and it regains its former speed on emerging into air.

It always depends on how far a wave is from its center of curvature what amount of curve it will have. Study the following examples:

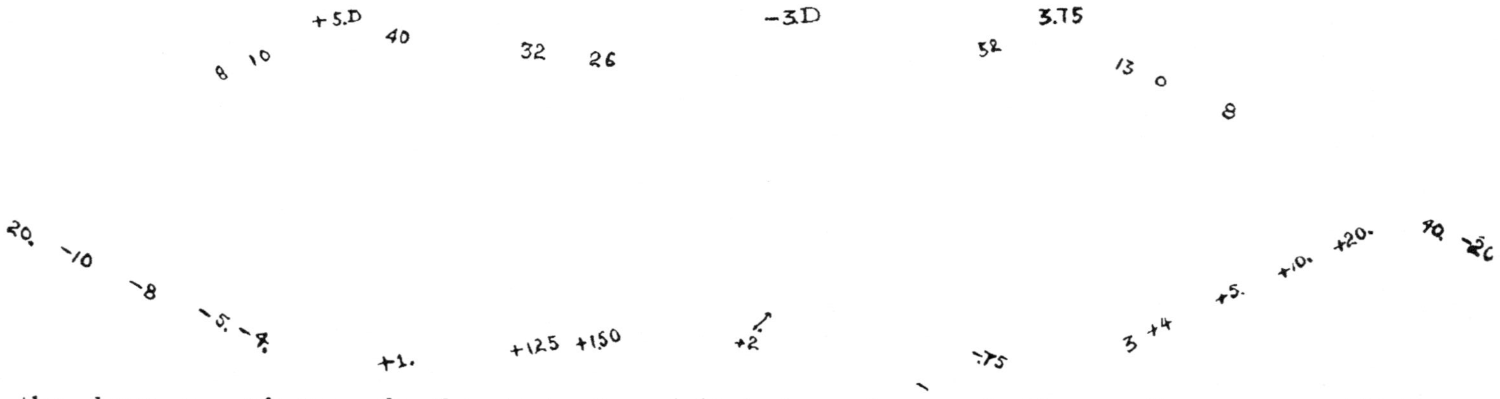

8 10 +5.D 40 32 26 —3D 52 3.75 13 0 8

20. —10 —8 —5. —8. +1. +125 +150 +2 .75 3 +4 +5. +10. +20. 40 20

In the bove : pie we ahe the cand e p ced 10 inches fom a +5 sphere, ten a space of 20 in. to —3 D. sphere. Again a sp ca of 12 inches to the +3.75 D. Bow as all waves of l ght tat are leaving a po are ins, the light wi l enter the first lens with a —4 curve and emerge from it wth a +1 curve, wh wo fous at 40 inches, providing it was not interrupted; bt as the —3 D. lens is only 20 in. away, the light wil en it with a +2 rue and emerge from it a —1, wh we will se has me from a point 40 in. tht. has ten a ade of 12 in. before entering the ext lens, wh wld mean tat t wu d be 52 in. fr its rer of curvature as it enters the +3.75 D. It wld en er a —75 and emerge a +3 D. and fous at 13 The rays wd ten cross and begin to diverge, or in otr words, pass from + to — ew. The figu above the drawing represent inches from the rer ue, nd those below the ture of te wa of light.

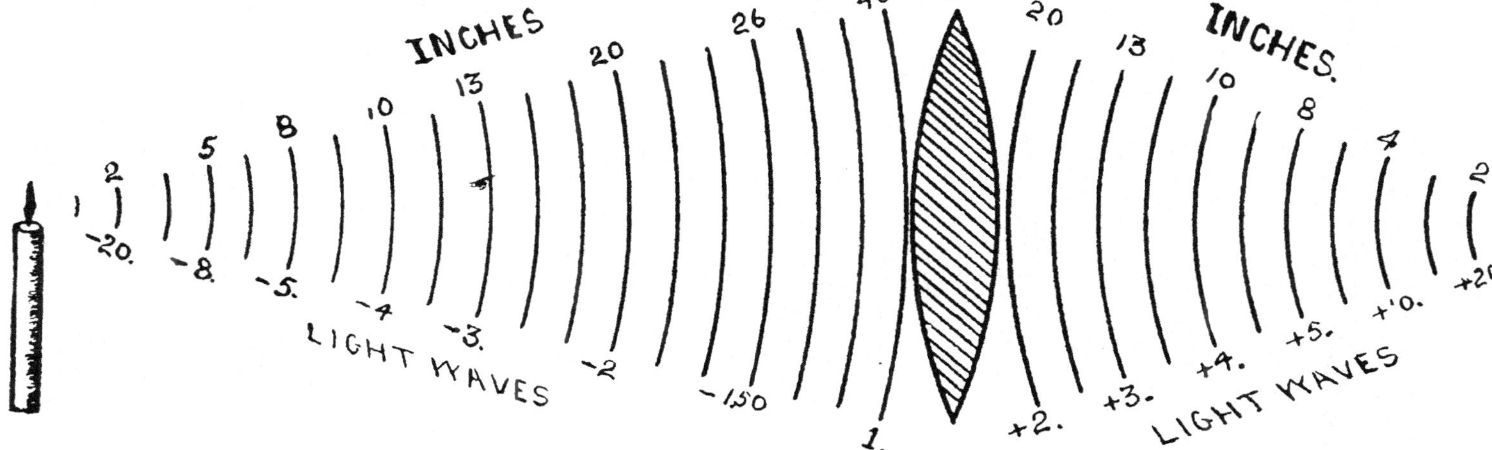

In h s examp e we have what s known as conjugate foci, two focal poin s at ached o one another

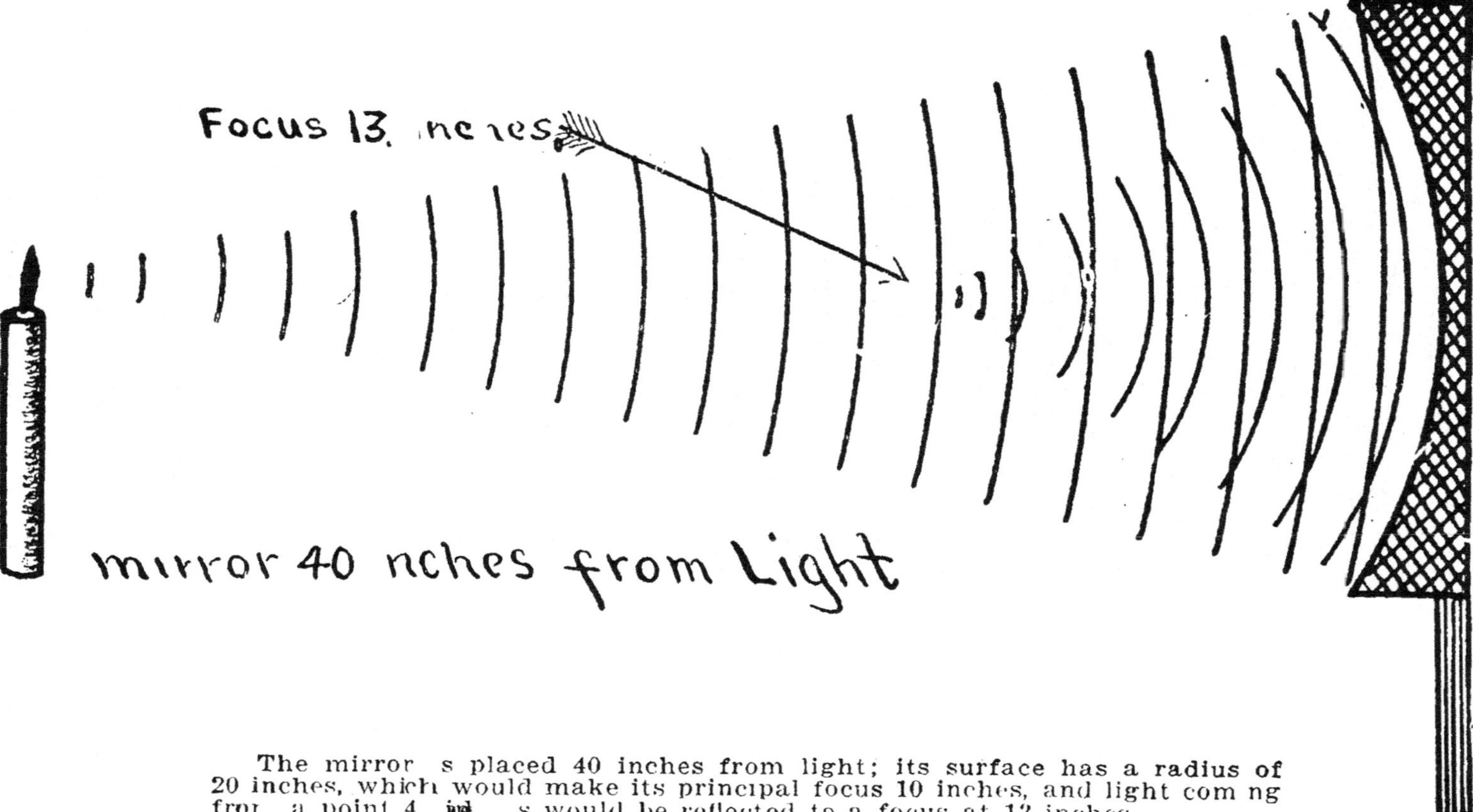

+4. Catoptry Mirror.

The mirror s placed 40 inches from light; its surface has a radius of 20 inches, which would make its principal focus 10 inches, and light com ng frm a point 4 s would be reflected to a focus at 13 inches.

Light, Recomposition of. The reuniting of the colors of the spectrum so as to produce white light. It is done by placing a second prism exactly like the first, with its apex turned in the opposite direction. The light will be recomposed and will emerge from the second prism as white light.

Light Area on the Face. The term used to designate the light upon the face when the beam of light from the retinoscope is directed upon the eye under observation.

Light Area in Pupil. The light seen in the pupil of an eye under observation with the retinoscope, caused by the reflex from the retina. Its character and relative movement indicate the refractive condition of the eye.

Limbus (L. border). Edge; border; hem.

Limbus Cornea (lim′-bus). (Border line.) The region where the cornea and sclerotic join.

Limit Angle. See Critical Angle.

Limitans (lim′-it-ans). (L. limitare = to limit.) That which limits or bounds a body or organ.

Line of Fixation. A line which connects the object looked at with the macula lutea through the nodal point of the eye.

Line of Vision. The line which connects the object looked at with the fovea centralis (visual axis).

Lippitudo (lip-pi-tu′-do). (L. lippus = blear-eyed.) An inflammation of the margins of the eyelids.

Liquor Morgagni. A small quantity of fluid between the lens and its capsule.

Logadectomy (log-ad-ek′-to-my). A removal of a part of the conjunctiva by means of a sharp knife.

Logades (log′-ad-ees). (Gr. logades = the whites of the eyes.) The first coat or tunic of the eye.

Logarithm (log′-a-rithm). (Gr. logos = word + arithmos = number.) The exponent of the power to which a number called the base (in the common system, 10) must be raised to produce a number

Loimophthal′mia. Contagious ophthalmia.

Longitude (lon′-ji-tud). (L. longus = long.) The angle at the pole between two meridians, one of which, the Prime Meridian, passes through some conventional point and from which the angle is measured.

Long-Sightedness. See **Hyperopia.**

Lorgnette (lorn-yet′). Double eye-glasses attached to a handle. This term is often applied to opera-glasses.

Louchettes. A kind of opaque glasses in which, for each eye, there is a small opening which makes it impossible to look in any other way than through this opening.

Loxophthalmos (lok-sof′-thal-mus). (Gr. loxos = slanting + ophthalmos = eye.) That condition in which the eye is turned from parallelism. (Strabismus; **H**eterotropia.)

Lucid (L. lucere = to shine). Clear, distinct.

Lucifugal (lu-sif′-u-gal). That condition which exists where a person avoids bright light.

Luminous Bodies. Those sources of direct light, as the sun, a lighted candle, etc.

Luminous Pupil. The appearance of the pupil under observation with the retinoscope.

Luxation (luk-sa′-shun) **of Lens** (L. luxare = to

dislocate). That condition where the crystalline lens is dislocated.

Lymph (L. lympha = water). A transparent, slightly yellow liquid which fills the lymphatic vessels. It is occasionally rose color from the presence of blood corpuscles, and is often opalescent from particles of fat. **L. System** is a collective term for the lymphatic glands and vessels.

Lymphatic (lim-fat′-ik). Pertaining to, of the nature of, containing, or conveying lymph.

MACROPSIA (mak-rop′-si-ah). (Gr. makros = large + opsis = vision.) That condition of an eye in which objects appear larger than they really are.

Macroscopic (mak-ro-scop′-ic). Gr. makros = large + skopeo = I view.) That which may be seen with the naked eye.

Macula Lutea (mak′-yu-lah lew′-te-ah). (L. a spot + lutea = yellow.) The small yellow spot of the retina which lies directly in the visual axis. It is about 4 mm. to the temporal side of the center of the optic disc, on the horizontal meridian, and is less than 2 mm. in diameter. A depression in its center, fovea centralis, is the most sensitive point of vision.

Madarosis (mad-ar-o′-sis). (Gr. "bald.") That condition in which the eyelashes are permanently destroyed.

Maddox Rod (Ernest E. Maddox, English ophthalmologist). An opaque disc with a slit through the center. Over this slit is placed a glass rod or cylinder. In looking through this rod at a small round light it causes the light to look like

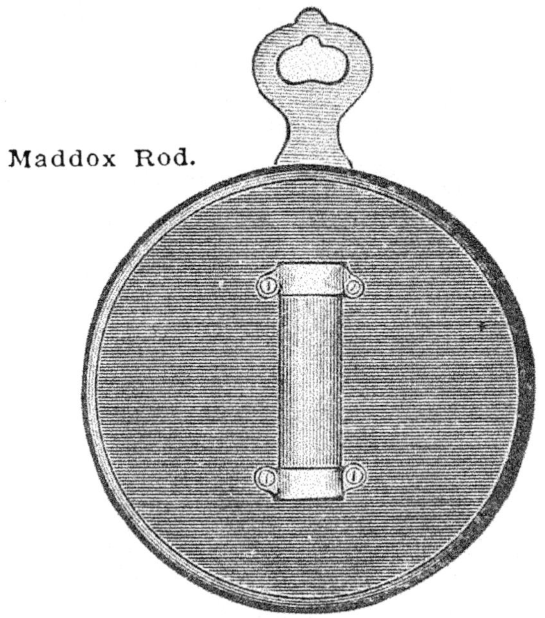
Maddox Rod.

a long streak. This rod with a red glass disc over the other eye is used for testing muscular imbalance. Before testing for muscular imbalance correct all errors of refraction.

Madisterium (mad-is-ter'-i-um). An instrument used for removing the eyelashes.

Magnet Operation. A method used for removing particles of iron and steel embedded within the tissue of the eye, by means of a magnet.

Magnify (mag'-ni-fy). To render an apparent increase in the size of an object above that seen in emmetropia.

Malacia (mal-a'-se-ah). (Gr. malakia = a softness.) Morbid softening of tissue.

Malacocataracta (mal-ak-o-kat-ar-ak'-tah). A soft cataract which forms in the crystalline lens of a person under the fortieth year. This form of cataract is usually the result of injury.

Malaxation (mal-ax-a′-shun). (L. malaxare = to soften.) A rubbing or kneading of the eyeball.

Malignant (mal-ig′-nant). (L. malus = evil.) "Destructive, to cause distress." A term applied to any disease whose symptoms are such as to threaten the destruction of the part. See Myopia.

Malingerer (Fr. malingre = sickly). One who pretends to have a defect of vision or some other function, to evade duty.

Marginal Keratitis (mar′-jin-al ker-at-i′-tis). (L. margo = a border.) Relating to a border or edge. A disease of an inflammatory nature which occurs usually in elderly people. The inflammation extends around the rim of the cornea. If the process continues the cornea is invaded by a densely vascular, superficially ulcerated, and yet thickened zone.

Marmarygea (mar-mar′-ij-e-a). (Gr. "brilliant.") Appearance of sparks before the eyes.

Mass. (Physics) The quantity of matter which a body contains, irrespective of its bulk or volume.

Mature (ma-tur′). (L. maturare = to ripen.) Fully developed; ripe.

Means (menz). The second and third terms of a proportion.

Measure (mezh′-ur). (L. mensura = measure.) A unit or standard adopted to determine the length, volume, or other quantity of some other object.

Media (L. medius = middle). Intervening bodies, such as glass, air, water, etc., which light can pass through freely. Media is plural of medium.

Median line, n. A line perpendicular to and bisecting the distance between the centers of rotation of the eyes.

Medium. Intervening body or quantity. The dioptric media of the eye consist of the cornea, aqueous humor, crystalline lens, and vitreous humor.

Megalocornea (meg-al-o-kor'-ne-ah). (Gr. megas = large.) That condition in which there is bulging of the cornea.

Megalopia (meg-al-o'-pi-ah). (Gr. megas = great + ops = eye.) See Macropsia.

Megalopsia (meg-al-op'-si-ah). (Gr. megas = great + opsis = vision.) That condition of the eye in which objects appear larger than they really are.

Meg'aloscope (Gr. megas = great + skopeo = I view). A large magnifying lens.

Megascope (meg'-a-skope). A microscope for examining large objects.

Megophthalmus (meg-of-thal'-mus). (Gr. megas = great + ophthalmos = eye.) That condition in which the part of eye is abnormally large.

Meibomian Glands (mi-bo'-me-an; after Meibomius, the discoverer). A variety of glands which are embedded in the tarsal cartilages. There are from thirty to forty in the upper lid and from twenty to thirty in the lower lid. Their ducts open upon the free margin of the lids. These glands secrete a sebaceous, oily fluid which assists in lubricating the lids as they glide over the eyeball, and also prevents the lids from sticking together when we have them closed. Another function is, that as the margins of the lids are kept oily at all times the tears do not flow over them so easily. This oily substance also mixes with the tears and assists

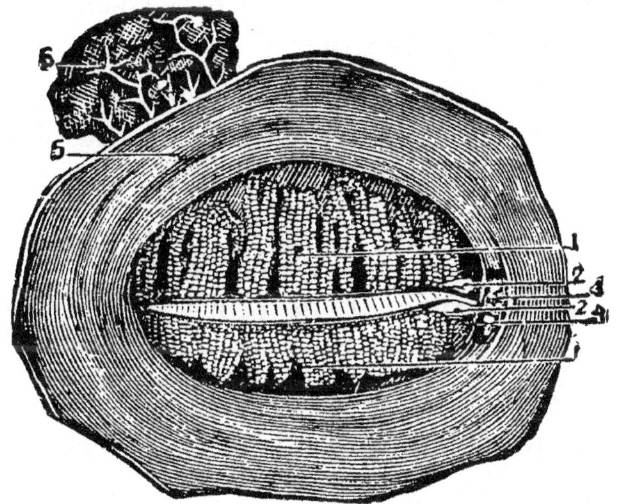

ANATOMY OF LIDS.

No. 1. Meibomian Glands.
No. 2. Puncta Lachrymalia.
Nos. 3 and 4. Lachrymal Canals.
No. 5. Orbicularis Muscles.
No. 6. Lachrymal Glands.

in preventing friction between the eyeball and lids, and at the same time prevents the cornea from becoming dry so quickly. Also known as tarsal glands. See Chalazion.

Mel'anin (Gr. melas = black). A dark pigment from choroid, hair, and other dark tissues.

Melasma (Gr. melas = black). Melanoderma; morbid cutaneous discoloration.

Melasma Palpebrarum (mel-as'-ma pal-pe-bra'-rum). A discoloration of the eyelid, which occurs frequently in pregnant women.

Membrana Capsularis. That portion of the arteria centralis which forms a vascular network and coats the posterior surface of the lens.

Membrana Pupillaris (mem-bra′-na). A membrane covering the pupil of fetal life. This sometimes fails to disappear up to time of birth.

Membrane (mem′-bran). A thin tissue covering some surface or organ.

Membrane Nictitating (L. nictitare = to wink). That which is sometimes called the third eyelid, to be seen in various animals.

Meniscus (men-is′-kus). (Gr. meniskos = crescent.) A crescent. When applied to a lens, would mean that it is more periscopic than necessary to produce the same refracting power.

Menotyphlosis (men-o-tyf-lo′-sis). A condition of the eye in which there is diminution of vision during night.

Mensuration (men-su-ra′-shun). The science of measuring.

Meramaurosis (mer-am-au-ro′-sis). A condition of the eye in which part of the field of vision is lost; partial amaurosis.

Meridian (mer-id′-i-an). In optics, any straight line passing from edge to edge on the surface of a lens and over its optical center. A meridian of the cornea is any straight line crossing its surface through its anterior pole. A circle describes 360°, one-half of which, or 180°, is marked on the trial frame. 0 and 180 are found in the horizontal, 90° describes the vertical. On the clock-dial (astigmatic chart) they are numbered from left to right, which must be remembered when fitting glasses, and on trial frame from right to left. These figures on trial frame correspond with clock-dial when they face each other.

Meropia (mer-o′-pi-ah). (Gr. part, vision.) See Amblyopia.

Mesoropter (mes-o-rop′-ter). The position of eyes in state of absolute rest.

Mesoseme (mes′-o-sem). (Gr. mesos = middle + sema = sign.) Having a medium orbital index between 84 and 89.

Mesoretina (mes-o-ret′-in-a). (Gr. mesos = middle.) The middle layer of the retina.

Metamorphopsia (met-am-or-fop′-si-ah). (Gr. meta = over + morphe = shape + opsis = vision.) That condition of the eye in which objects appear distorted.

Metastasis (met-as′-ta-sis). The shifting of the seat of a disease from one part of the body to another.

Meter (me′-ter). (Gr. metron = measure.) Unit of length in the metric system; the equivalent of 39.371 inches.

Meter Angle. The angle through which each eye turns when it abandons parallelism of the binocular lines (orthophoria) of fixation in order to fix an object situated upon the median line one meter from the eye. It is the arc embraced between the median line and the line of fixation, whose length is one meter. The average distance between pupils in the adult is 62 mm.; then one-meter angle equals a deviation for one eye of 31 mm. As the eyes converge closer the meter angles increase in proportion. This description refers to perfect muscular balance only. The writer differs from other authors, inasmuch as he claims that if the eyes were diverging one meter angle from parallelism, when in a state of rest, they will be forced to use two meter angles of convergence in order to view an object one meter distant. This description applies to the different varieties of muscular imbalance.

Meter Lens. A lens that will focus parallel rays of light at a distance of one meter.

Metric Curve. A curve that has a radius of one meter; two M, C. has a radius of 20 inches.

Metric System. A decimal system of weights and measures adopted in France (1795) and various European countries, and gradually coming into general use for scientific purposes. Ten units of one denomination marking one unit of the next higher. A meter is equivalent to about 39.37 English inches.

Decimal parts of the standard or principal unit are denoted by Latin prefixes; multiples of the same, by Greek prefixes.

From the Latin			From the Greek		
Milli	means	0.001	Myria	means	10,000
Centi	means	0.01	Kilo	means	1,000
Deci	means	0.1	Hekto	means	100
			Deka	means	10

In the tables units in common use are in **bold-faced type.**

Units of Length

Standard unit, the **Meter**

Table

10 **millimeters** (mm.)	=	1 **centimeter** (cm.)
10 centimeters	=	1 decimeter (dm.)
10 decimeters	=	1 **meter** (m.)

10 meters	= 1 dekameter (Dm.)
10 dekameters	= 1 hektometer (Hm.)
10 hektometers	= 1 kilometer (Km.)
10 kilometers	= 1 myriameter (Mm.)

Microblepharia (mi-kro-blef-a'-ri-ah). (Gr. mikros = small + blepharon = eyelid.) A very narrow and thin eyelid.

Microcornia (mi-kro-kor'-ne-ah). (Gr. mikros = small.) A small cornea.

Microlentia (mi-kro-len'-ti-ah). A very small crystalline lens.

Micrometer (Gr. mikros = small + metron = measure). An instrument which is used for making measurements of very small bodies.

Microphthalmia (my-krof-thal'-mee-ah). (Gr. mikros = small + ophthalmos = eye.) Abnormally small eyes.

Micropsia (my-krop'-see-ah). (Gr. mikros = small + opsis = sight.) That condition of the eye in which objects appear smaller than they really should.

Microscope (mi'-kro-scope). (Gr. mikros = small + skopeo = I view.) An optical instrument used for examining minute objects.

Migraine (mi-gran'). (Gr. hemi = half + kranion = skull.) A kind of sickness or nervous headache, usually periodical and confined to the side of the head.

Mikron (mi'-kron). (Gr. mikros = small). Millionth part of a meter.

Milium (L. millet = seed). A small elevation, on the skin of the eyelid, filled with a greasy secretion.

Milphae (mil′-phae). A morbid condition in which the eyelashes drop out.

Milphosis (mil-fo′-sis). That condition in which the eyebrows as well as the eyelashes have fallen out.

Minify (min′-i-fy). To render an apparent decrease in the size of an object from that seen in emmetropia.

Minuend (min′-u-end). (L. minuere = to lessen.) The number from which another number is subtracted.

Minus (mi′-nus). Less; that from which something has been subtracted; negative; as a minus quantity. Minus lenses, same as concave.

Miosis (mi-o′-sis). (Gr. meiosis = a lessening.) Excessive contraction of the pupil.

Miotic (mi-ot′-ic). Any agent or medicine which causes the pupil to contract.

Molecule (mol′-e-kul). The smallest particle of any substance.

Molluscum (mol-us′-kum). (L. mollis = soft.) A skin disease with pulpy tumors, chiefly on the face, neck, and trunk, due to altered gland secretion. A small pearly, warty elevation which breaks down in the center and discharges a cheeselike material. It is said to be contagious

Momentum (mo-men′-tum). (L. momentum = motion.) The quantity of motion in a moving body, being always proportioned to the quantity of matter multiplied into the velocity.

Monoblepsis (mon-o-blep′-sis). (Gr. monos = one + blepsis = sight.) That condition in which objects are seen distinctly only when the eyes are used singly, vision being imperfect when both eyes are used.

Monochromatic Light (Gr. monos = single + chroma = color). The spectrum is formed by a prism dividing light into its seven colors. Such light is called Monochromatic Light.

Monocle (mon'-o-kl). An eyeglass to be worn over one eye only.

Monocular (mon-ok'-u-lar). (Gr. monos = single + L. oculus = eye.) Having one eye only. Monocular vision is that condition where the patient has vision in one eye only

Monoculus. A monster with one eye only.

Moon-Blindness. Amblyopia caused by having the eyes exposed to the full glare of the moon for considerable time.

Monoparesis (mon-o-par'-es-is). (Gr. monos = single + paresis.) Paralysis of a single part of the body.

Monopathy (mon-op'-ath-e). (Gr. monos = single + pathos = suffering.) A local disease of one organ or function.

Monops (mon'-ops). (Gr. monos = single + ops = eye.) A foetus with but one eye.

Motor (L. "a mover"). A nerve center controlling motion. A muscle causing motion.

Motor Muscles. The muscles that control the movements of, or parts of, the eyes—the recti, the oblique muscles, the ciliary, and the iris muscles.

Mucocele (mu'-ko-sele). (L. mucus + Gr. kele = tumor.) Distention of the lachrymal sac, chronic thickening of the lining membrane, and increased secretion of mucus. The mucus may be clear or turbid. Any tumor containing mucus.

Muscae Volitantes (mus′-cae vol-i-tan′-tes). (L. "flying flies.") Small floating bodies, resembling sticks, etc., which move about in the field of vision, but do not actually cross the fixation point, and never interfere with sight. They are usually seen against some bright object. They depend upon minute changes in the vitreous, which are present in nearly all eyes. They vary, or seem to vary, greatly with the health and state of the circulation, but are of no real importance. They are most abundant and troublesome in myopic eyes.

Muscle (L. musculus = a little mouse, from the resemblance of a muscle in contraction to the movements of a mouse under a cloth). Muscles. (Eye) The eyeball is rotated around its center of rotation by the individual or combined action of six muscles, namely, four recti or straight, and the two oblique; the seventh muscle, levator palpebrae, is attached to the upper lid, which it raises.

Of the six muscles inserted into the eyeball, five take their origin from the apex of the orbit, while the sixth and shortest, the inferior oblique, takes its origin from the superior maxillary bone at the anterior and nasal side of the orbit.

Internal Rectus Muscle turns the eye in, and is supplied by the third cranial or motor oculi nerve.

Superior Rectus Muscle turns the eye up, and

is supplied by the third cranial or motor oculi nerve.

Inferior Rectus Muscle turns the eye down, and is supplied by the third cranial or motor oculi nerve.

Inferior Oblique Muscle rolls the eye on its optic axis, drawing the bottom and back part of the eye in and down while the front moves up and out, and is supplied by the third cranial or motor oculi nerve.

External Rectus Muscle turns the eye out, and is supplied by the sixth cranial or abducens nerve.

Superior Oblique Muscle rolls the eye on its optic axis, turning the back part of the eye inward and upward while the front part moves down and out, and is supplied by the fourth cranial or patheticus nerve.

Ciliary Muscles are inside the eyeballs, and are used for accommodating only. They are supplied by the third nerve.

Orbicularis Palpebrarum Muscle closes the lids, and is supplied by the seventh, or facial nerve. (One of the muscles of expression.)

Levator Palpebra Superioris lifts the lids, and is fed by a branch of the third cranial nerve.

Sphincter Muscle, which closes the pupil, is supplied by the third nerve.

Radiating Muscles of the iris, which dilate the pupil, are supplied by the sympathetic nerve.

Horner's Muscle. See Tensor Tarsi.

Palpebral Muscles are two involuntary muscles, superior and inferior. The superior is principally inserted into the upper margin of the tarsal plate. The inferior, found in the lower lid, interwoven with the fasciculi of the orbicularis palpebrarum.

Riolan's Muscle is the muscular tissue surrounding the hair follicles and glands of Moll, near the border of the eyelids.

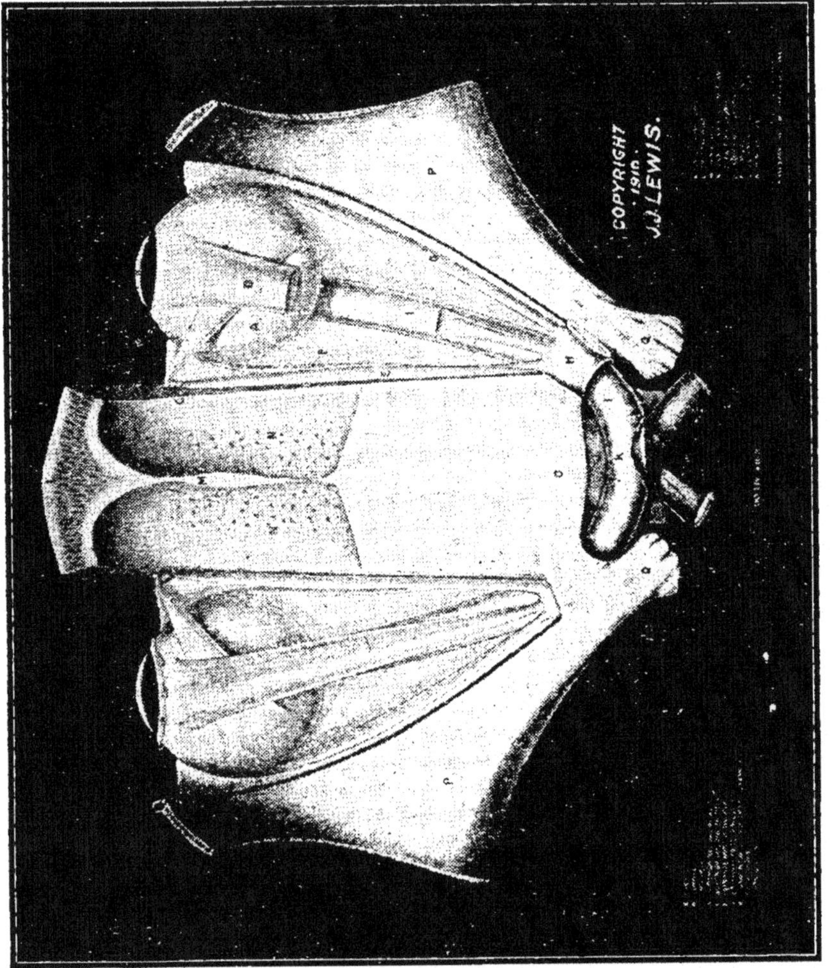

Muscles controlling movements of Eyes.

Tensor Tarsi compresses the lachrymal sac and pulls the puncta against the eyeball. Supplied by the seventh cranial nerve.

Corrugator Supercilii draws eyebrow down

and inward, and is supplied by the seventh facial nerve.

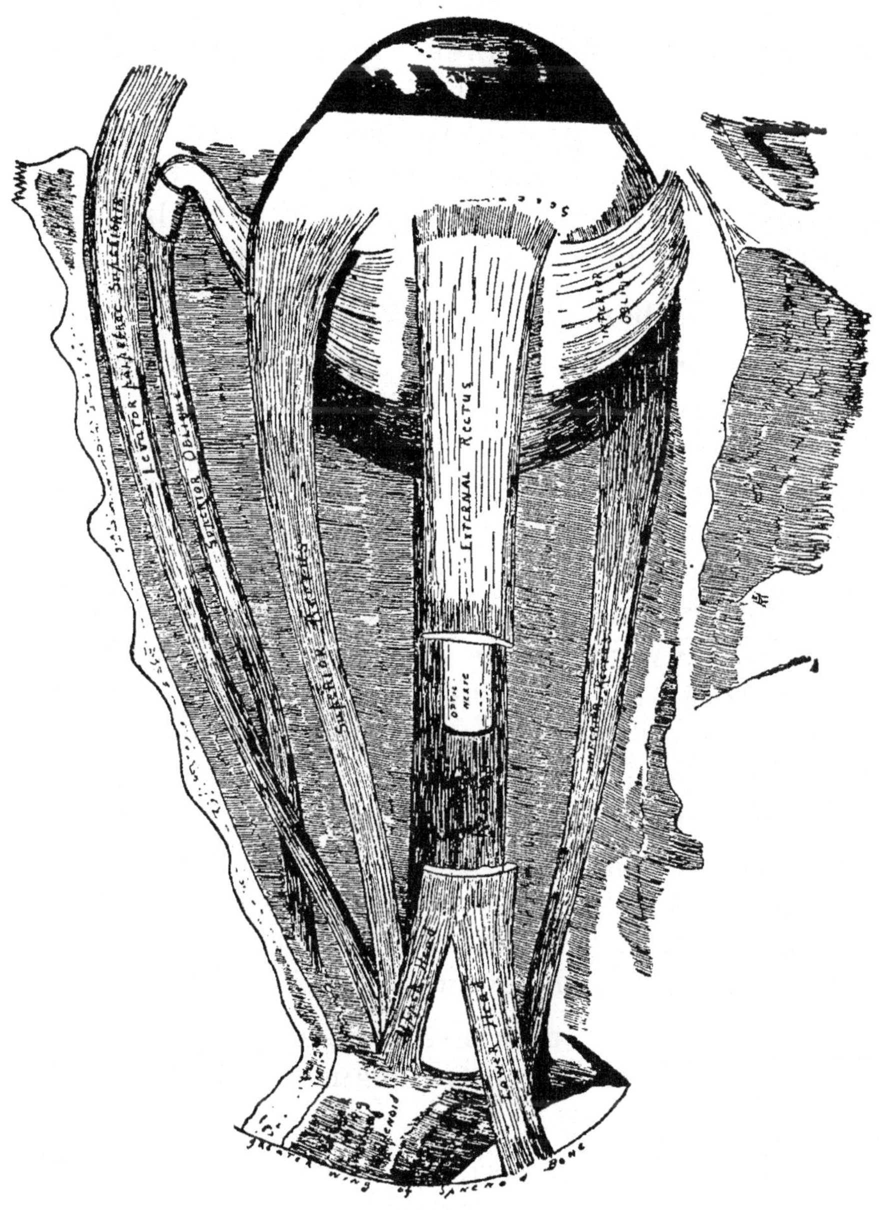

Cut showing the Extrinsic Muscle.

Pyramidalis nasi. This muscle depresses the eyebrow. Supplied by the seventh facial nerve.

Muller's Muscle. Bands of circular fibers situated internal to the radiating muscles in the ciliary body. They are sometimes called the "ring muscle" of Muller. **Fibers of Muller** are the radiating fibers which pass through nearly the entire thickness of the retina, supporting its different layers and binding them together. They form at one end the membrana limitans interna and at the other end the externa.

Muscular Asthenopia. See Asthenopia.

Muscular Imbalance. It is generally agreed by the authorities of today that ametropia is responsible for 90 per cent of Muscular Imbalance. For this reason it is considered advisable to always correct any ametropia that may be present, and have the patient wear the correction for at least six weeks. At the end of this time, should any muscular imbalance be manifest, correct half of the amount. Always test for muscle trouble while the patient is wearing his full correction for the ametropia, otherwise it will not be considered a proper test. There are but few exceptions to this rule: e. g., when a prism, base in, will allow you to decrease a minus lens or increase a plus, prescribe it. See Heterophoria and Heterotropia.

Mycophthalmia (mi-kof-thal′-mi-ah). Inflammation of the conjunctiva, caused by a spongy growth.

Mydriasis (mid-ri′-as-is). Dilatation of the pupil, caused by the use of atropine or other mydriatics, or paralysis of the motor oculi nerve.

Mydriatic (mid-ri-at′-ic). A drug that dilates the pupil.

Myiocephalon (my-i-o-sef′-al-on). (Gr. myia = fly + kephale = head.) A small protrusion of the iris through a perforation of the cornea.

Myitis (mi-i′-tis). (Gr. mys = muscle + itis = inflammation.) Inflammation of the muscles.

Myodesopsia (Gr. myia = fly + opsis = sight). See Muscae Volitantes.

Myograph (my′-o-graph). (Gr. mys = muscle + grapho = I register.) An instrument for recording the different phases, such as the velocity, intensity, etc., of a muscular contraction, with the aid of a registering apparatus.

Myography (my-og′-ra-phy). (Gr. mys = muscle + graphe = a drawing.) A description of muscles, including the study of muscular contraction, with the aid of a registering apparatus.

Myology (my-ol′-o-ji). (Gr. mys = muscle + logia = a discourse.) A description of the muscles of the human body.

Myologist (my-ol′-o-gist). One skilled in that part of anatomy which treats of muscles.

Myope (my′-ope). A near-sighted person.

Myopia (my-o′-pi-ah). (Gr. myo = close + ops = eye.) An optical defect of the eye which causes parallel rays of light to focus in front of the retina, with the muscles of accommodation at rest. It is also known as Brachymetropia and Hypometropia. A correction (concave lens) is necessary before normal vision can be obtained. In the early days the Greeks noticed that all near-sighted persons could see better at a distance by half closing the eyelids (squinting), hence the name myopia. The eyeball may be too long (axial) or the refraction too great, causing the parallel rays to cross and meet the retina as divergent rays, which form a circle of diffusion, and so cause a blurred and indistinct image of the object. **Myopia** from excess of curvature is much rarer than the axial form. We sometimes see a case of apparent myopia due to excess of curvature of the lens, caused by a spasm of the ciliary muscle. This is what is known as false myopia, and will disappear under the influence of atropine, or rest.

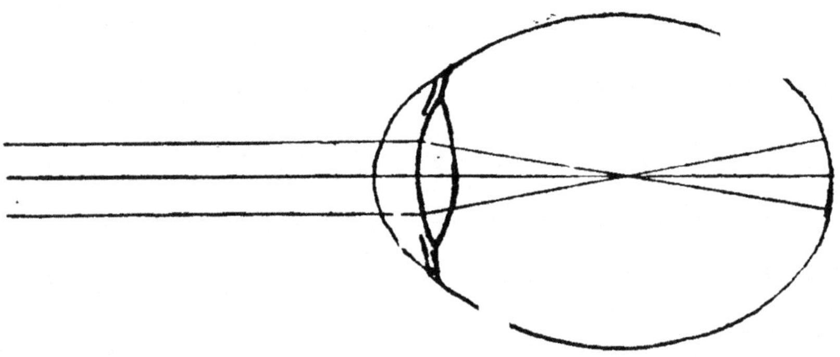

These cuts have not been made with mathematical precision, but are merely intended to roughly exemplify the principle involved. The above illustrates the effect of a distant point upon a myopic eye. It will be observed that the rays from the distant point cause the image from that point to spread out over a considerable area on the retina. A million points would make a million blurred spots. They would overlap each other and render the picture indistinct.

Donders pronounced every highly myopic eye a diseased eye, but of late years it has been shown that this remark is liable to convey a false meaning. It is admitted that up to 3-D. the cases are seldom serious, it being generally possible to give perfect vision by proper glasses, but when the case is above 5-D. we do not always obtain perfect vision by proper lenses. and look for various pathological conditions.

Symptoms: The patient sees distant objects badly and near objects better. The pupils are usually large, and as presbyopia advances they contract. The ciliary muscles are smaller and weaker than in the normal or emmetropic eye. As a rule, myopic patients cannot wear their full correction when first fitted, but after wearing about two-thirds of the correction for about six weeks the full amount may be prescribed. Myopia that is gradually on the increase is called Progressive Myopia. Myopia that is of a rapidly progressive type, and is very destructive to the tissues of the eye, is called Malignant Myopia.

Myopic Crescent. As seen by the ophthalmoscope, is a white crescent at the outer side of the optic disc. This condition is caused by the choroid being torn away from the optic disc and allowing the sclerotic to show through. Found in high degrees of myopia.

Myosis (my-o′-sis). (Gr. myo = I close the eye.) Abnormal contraction of the pupil. (Same as Miosis.)

Myositis (my′-o-si′-tis). (Gr. mys = muscle + itis = inflammation.) Inflammation of the muscles.

Myotic. (Gr. myo = I close the eye.) An agent that will contract the pupil, such as eserin, pilocarpine, etc.

Myotomy (mi-ot′-o-me). (Gr. mys = muscle + tome = a cutting.) The dissection or division of muscles.

NASAL DUCT (L. nasus = nose + ducere = to lead). That part of the tear duct which extends from the lower part of the lacrimal sac to the inferior canal of the nose. It is about five-eighths of an inch in length and lined with a mucous membrane similar to the lacrimal canals. At the end of the duct we find the valve of **Hasner**.

Near Point (or **Punctum Proximum**). The nearest point at which the eye can see distinctly when employing its full amount of accommodation. It varies with the amount of accommodation the eye possesses. The way to determine the near point is to note the shortest distance at which an emmetrope can read small print with each eye separately. Properly speaking, the near point is that point for which the eyes' refraction is adjusted when the full amount of accommodation is being used.

Near-Sight. See Myopia.

Neb′ula (L. "cloud"). Slight corneal opacity.

Needling (need′-ling). An operation for soft cataract. The lens capsule is needled, and the aqueous allowed to absorb the lens.

Negative (neg′-a-tive). (L. negare = to deny.) The opposite of positive. The negative surface of a periscopic lens is the concave surface.

Neonatus (ne-on-a′-tus). (Gr. neos = new + L. natus = born.) Newly born.

Neotocophthalmia (ne-ot-ok-of-thal′-mi-ah). See Ophthalmia Neonatorum.

Nephablepsia (nef-ab-lep′-si-ah). See Snow-Blindness.

Nephelopia (nef-el-o′-pi-ah). (Gr. nephele = cloud + ops = eye.) A diminution of vision, caused by a cloudiness of the transparent parts of the eye.

Nephritic Retinitis (nee-frit′-ik). A form of inflammation of the retina associated with Bright's disease of the kidneys, characterized by white streaks along the course of the blood-vessels.

Nerve (L. nervus = nerve). A white string-like fiber which transmits impressions from an organ to the brain or from the brain to an organ. **Cranial N.**, any nerve arising from the brain direct. There are twelve cranial nerves, as follows:

1. Olfactory, special sense of smell.
2. Optic, special sense of sight (retina).
3. Motor Oculi, motor nerve for eye muscles.
4. Patheticus, motor nerve for superior oblique muscle.
5. Trigeminus, sensory, motion, and taste.
6. Abducens, motor nerve for external rectus muscle.
7. Facial, motor nerve for muscles of face.
8. Auditory, special sense of hearing.
9. Glosso-Pharyngeal, sensation and taste.
10. Pneumogastric, sensation and motion.
11. Spinal Accessory, motion.
12. Hypoglossal, motor nerve of tongue.

The nervous system is a system of connection and communication by which the different organs, vessels, and various parts of the body are brought into direct relation with each other and with the mind, and the various organs stimulated to harmonious or alternating action. It consists of the brain and spinal cord, called the central nervous system, which controls the voluntary actions of the body, sometimes called the nerves of animal life, and is directly connected with the sympathetic nerves, which have been termed nerves of organic life, they being involuntary nerves and control the involuntary action of the various vital processes of the body. The nervous system is divided into the cerebro-spinal or central, sympathetic, and the vaso motor. The vaso motor system is a part of the sympathetic system and consists of the vaso motor center located in the medulla oblongata; of certain other subsidiary vaso motor centers in the spinal cord, and of vaso motor nerves. This system is connected with the blood-vessels in the various parts of the body, the muscular coats of which are supplied with filaments and plexuses of vaso motor nerves which regulate the size of the blood-vessels. They are of two kinds: vaso dilators, stimulation of which causes dilatation of the blood-vessels and an increased amount of blood to a part, and vaso constrictors, stimulation of which causes constriction or contraction of the blood-vessels and a diminished amount of blood to a part. This last named system is very important to the practitioners of manipulatory forms of healing, and has only in the past few years been known to any extent, the vaso motor center being discovered by Schiff in 1855, and more accurately localized by Ludwig in 1871. The cranial nerves are those that have their apparent origin in the cranium. Sommering and other European anatomists name twelve pairs, while Willis and a few other authors designate only nine pairs, according to the order in which they pass out of the base of the brain.

Motor N., one which contains wholly motor fibers. **N. Center**, a group of cells which consist of gray matter and have a common function. **Mixed N.**, a nerve which is both motor and sensory. **N. Head**, the optic disc or papilla. **Sensory N.**, any nerve which transmits sensations or impulses. **Sympathetic N.**, any nerve of the sympathetic system.

Optic Nerve. The nerve which transmits retinal sensations from within the eye to the brain, and is known as the second cranial pair.

The Nerve of Vision. It is about 5 cm. in length and may be spoken of in three ways—the intracranial, the intraorbital, and the intraocular.

The optic nerves are noticeable for their size and their running a longer course within the cranium than within the orbit, and that they furnish no branches from their origin to their termination

Transverse sections of the optic nerve show it to be composed of about eight hundred distinct bundles of nerve fibers, separated from each other by connective tissue from the pia mater. The entire number of fibers contained within the optic nerve probably approaches a half million. These fibers after entering through the lamina cribrosa (sieve-like opening) again pass through the choroidal fissure (hole in choroid) and spread out in such a way as to form the shape of a wineglass (the retina). The fibers themselves are not sensitive to light, but each one terminates in a sensitive point (rods and cones) and in all parts of the third tunic (the retina), except at the entrance of the optic nerve (optic disc). See Retina; Optic Tract.

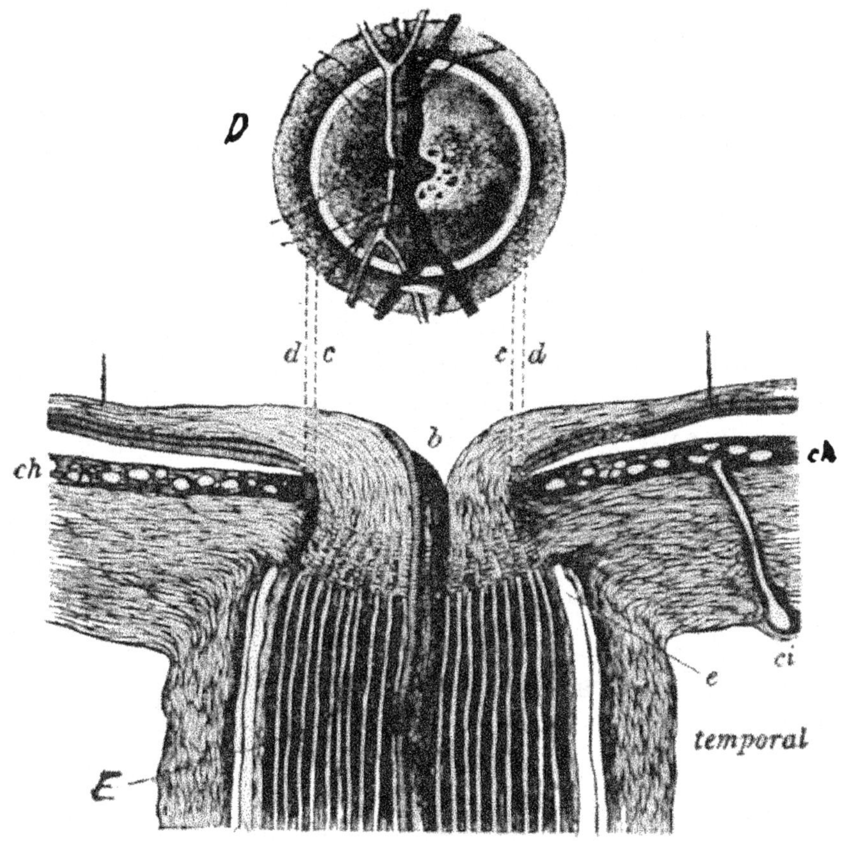

HEAD OF THE OPTIC NERVE.

D. Ophthalmoscopic view of the Optic Disc. The small excavations seen around the center is the Lamina Cribrosa. The Papilla is encircled by the white Scleral Ring (c) also the dark Choroidal Ring marked (d).
E. Longitudinal Section of Head of Optic Nerve; r. the Retina, b. Optic Excavation and Canal for Central Artery (porous opticus); ch. the Choroid; E. Optic Nerve, e. the narrow interspace which corresponds to the Scleral Ring seen by the Ophthalmoscope; s. Sclerotic; ci. Entrance of short Ciliary Artery.

Neurasthenia (nu-ras-then-i'-ah). (Gr. neuron = nerve + astheneia — weakness.) Exhaustion of nerve force.

Neuritis (neu-ri'-tis). (Gr. neuron = nerve + itis = inflammation.) Inflammation of the optic nerve.

Neurology (nu-rol'-o-je). (Gr. neuron = nerve + logia = discourse.) A study of the nervous system.

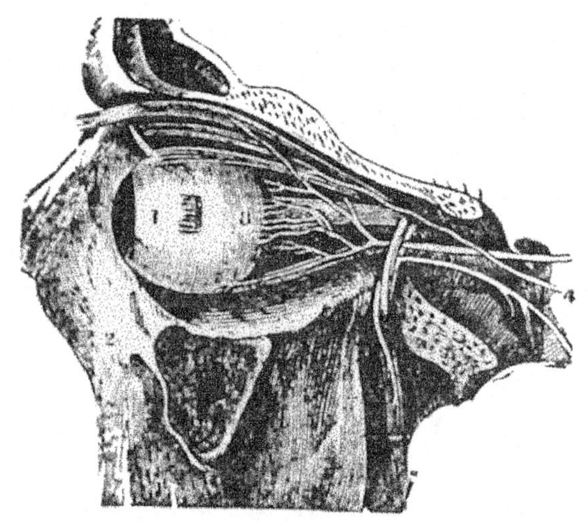

VIEW OF THE EYE FROM THE TEMPORAL SIDE WITH PART OF THE ORBIT REMOVED

1 Eyeball with part of external rectus muscle. 2 Superior Maxilla 3. Third pair (or Motor Oculi) nerves. They are distributed to all the muscles of the eye, except the Superior Oblique, External Rectus and the Dilator Muscles of the iris 4 Fourth pair of Nerves, feeding Superior Oblique Muscles 6 Sixth pair of Nerves, feeding External Rectus. 8 Cillary Nerves entering the globe.

Neurodealgia. Pain or excessive sensibility of the retina.

Neurodeatrophia. Atrophy of the retina.

Neuroretinitis (neu-ro-ret-in-i'-tis). (Gr. neuron nerve + L. retina + Gr. itis.) Inflammation of the optic nerve and retina.

Neutralize (neu'-tral-ize). (L. neuter = neither.) The method of counterbalancing or doing away with power in lenses. In order to determine whether a lens is of plus or minus power, hold it up a few inches from the eye and look at some distant object through it, then move the lens from side to side and if the object appears to move in the opposite direction to the movement of the lens it is a plus lens. If you wish to find its dioptric power, take from your trial case a minus lens and put them together and again look through them at the object, and should the object still move against the movement of the lenses the minus is not strong enough. On the other hand, should the movement be reversed and now the objects appear to move with the lenses the minus is too strong, and you must find the minus lens that will allow the object to remain stationary. Whatever minus lens is required to do this will be of the same power as your plus lens; for instance, it will require a — 3 sphere to neutralize a + 3 sphere. When you look at an object through a minus lens, and move the lens as explained above, the object will appear to move with the movement of your lens, and in order to find its dioptric power use plus lenses as in the previous test until all movement of the object looked at has disappeared, and then the minus lens will be the same power as your plus. If the lens is compound, use the weakest spherical lens which neutralizes the motion in one direction; this usually gives the spherical surface, then use a cylinder to neutralize motion at right angles to this. If in order to neutralize a given compound lens, + 2 sphere combined with + 1 cylinder, axis 90° is required, then the lens being neutralized is a — 2 sphere combined with — 1 cylinder, axis 90°, etc.

Nictitation (nik-tit-a'-shun). (L. nictitare = to

wink.) Involuntary convulsive twitching of the eyelids.

Night Blindness. See Nyctalopia.

Niphablepsia (nif-ab-lep′-si-ah). (Gr. nipha = snow + ablepsia = blindness.) That condition wherein blindness is caused by the glaring reflection of sunlight upon the snow. Snow-blindness.

No′dal Points, or Cardinal Points. Are two points situated on the optic axis, connecting the centers of curvature of the refracting compound dioptric system of the eye. The nodal points of the eye are so close together that they may be considered as one point.

Nodal Points of a Lens. The two points of the principal axis, so situated that every ray which, before being refracted, is directed toward the first of them, seems, after its refraction, to come from the second one, and takes a direction parallel to that which it had at first. These two parallel rays are called lines of direction, and act, in the combined system, the same part as the line passing through the nodal point of a single refracting surface.

Normal. (L. norma = rule.) That which conforms to the natural rule. A straight line drawn from any point of a curve or surface so as to be perpendicular to the curve or surface at the point which it strikes is said to be normal to the surface.

Normal Vision. Vision is said to be normal when an eye can read a line on Snellen's Test Type from the distance at which it is numbered. The smaller the objects that an eye can distinguish or the greater the distance at which it can distinguish an object of given size, the greater is the acuity of vision that it possesses. Suppose, for instance, that the eye is just able to distinguish the letters in the line marked 50 on the Snellen's test type from a distance of twenty feet, then the vision would be 20/50. The vision in this case would not be as good as if the line marked 40 had been read from the same distance; and in order to have normal vision the patient should read the line marked 20 at twenty feet, with each eye separately, then the vision would be known as 20/20. Sometimes the patient will read the line marked 15 or even 10 from twenty feet. In this case the vision is exceptionally acute, and is designated as 20/15 or 20/10.

Normal vision does not indicate that the eye is normal, as the patient may be straining to bring the vision up to this point, as in Facultative Hypermetropia. Again, an emmetropic eye does not always have normal vision.

Notation (no-ta'-shun). (L. notare = to mark.) A system of written signs of things and relations used in place of common language.

Nubecula (nu-bek'-u-la). (L. dim of nubes = cloud.) Slight cloudiness of the cornea.

Nuclear (nu'-kle-ar). Pertaining to the center. The controlling center of activity.

Nuclear Cataract. See Cataract.

Numeration. (L. numerare = to count.) The art of reading numbers.

Numerator (nu'-me-ra-tor). The number, in a

common fraction, which shows how many parts of a unit are taken.

Nyctalopia (nyk-tal-o′-pi-ah). (Gr. nycto = night + alaos = obscure + ops = eye.) Night blindness. A condition where a person does not possess normal night vision

Nyctotyphlosis (nyk-to-tyf-lo′-sis.) (Gr. nyx = night + typhlosis = blindness.) State of blindness at night time.

Nystagmus (nys-tag′-mus). (Gr. nystagmos = a nodding.) Short, jerking movements of the eye which are very rapidly repeated and always occur in the same direction. The movements of the eye, as a whole, are not affected by it. Defective vision of such cases is not to be attributed to the nystagmus, but, on the contrary, is the cause of it. **Vertical n.,** the eyes continually move vertically. **Lateral n.,** the eyes constantly move horizontally. **Rotary n.,** the eyes constantly rotate.

OBFUSCATION (ob-fus-ka′-shun). (L. ob = toward + fuscus = dark.) An obscuration of vision or a confusion of sight.

Object. Something visible or tangible. That which is seen. An external something the image of which is upon the retina, which is intelligently impressed and appreciated by the brain.

Objective (ob-jek′-tive). (L. ob = against + jacere = to cast.) Symptoms observed by operator usually with ophthalmoscope or retinoscope. Symptoms which the refractionist discovers by means of one or more of his five senses. **Objective Examination.** An examination conducted independent of the patient's statements; e. g., retinoscopy, ophthalmoscopy and many other tests by means of instruments.

Oblique. Slanting; placed in a plane between the horizontal and vertical planes.

Occipital (ok-sip'-it-al). Pertaining to the back part of head. **Occipital lobe** is the posterior portion of the cerebral hemisphere.

Occipito-Frontalis. The muscle which lifts the eyebrows upward. Supplied by the seventh nerve.

Occlusion of the Pupil (ok-klew'-shun). Blocking up of the pupil by a membrane.

Ocellus (o-sel'-lus). (L. "the eye.") A single eye.

Ocular (ok'-u-lar). (L. oculus = eye.) That which pertains to the eye.

Ocular Refraction. The science treating of the optical conditions of the eye, the estimation of its errors of refraction and their connection with lenses for the eye.

Ocular Spectres. Imaginary objects floating before the eyes.

Oculist (ok'-u-list). (L. oculus = eye.) A physician and surgeon who has received the degree, "Doctor of Medicine," and makes a specialty of the eye and its diseases.

Oculo-Motor Center is a point situated beneath the floor of the aqueduct of Silvius around which the impulse to use accommodation united with the action to use the different muscles of the eye is stimulated.

Oculomotor (ok′-yu-loh-moh′-tor). (L. oculus = eye + motus = motion.) Pertaining to the movements of the eye.

Oculus (ok′-yu-lus). The organ of vision.

O. D. Oculus Dexter. The right eye.

Offset Guard. An eye-glass guard with a long shank, the purpose of which is to hold lenses farther from the eyes.

Old Sight. See Presbyopia.

O′nyx. (Gr. "nail.") An accumulation of pus between the layers of the cornea, resembling a finger nail.

Opacity (o-pas′-i-ty). (L. opacus = obscure.) The quality of that which is opaque.

Opaque (o-pake′). Impervious to light. Not transparent.

Operculum. (L. "lid," cover.) Anything resembling a lid or cover.

Operculum Oculi (o-per-cu′-lum oc′-u-li). The eyelid.

Operation (op-er-a′-shun). (L. opus = work.) An act performed with instruments or by the hands of a surgeon.

Ophryitis (of-ry-i′-tis). (Gr. ophrys = eyebrow + itis = inflammation.) That condition in which the eyebrows are inflamed.

Ophrys (of′-rys). (Gr. eyebrow.)

Ophthalmagra (of-thal′-ma-grah). (Gr. ophthalmos = eye + agra = seizure.) A sudden intense pain in the eye, usually rheumatic or gouty in origin.

Ophthalmalgia (of-thal-mal′-ge-ah). (Gr. ophthalmos = eye + algos = pain.) Sudden violent pain in the eye, not the result of inflammation, but neuralgic in character.

Ophthalmatrophia (of-thal-mah-tro′-fe-ah). (Gr. ophthalmos = eye + atrophia = atrophy.) Atrophy of the eye.

Ophthalmia (of-thal′-mi-ah). Severe inflammation of the eye. This more particularly applies to the conjunctiva of the eyelids and eyeball.

Ophthal′mia Neonato′rum. (Gr. ophthalmos = eye + L. neonatus = new-born.) A form of purulent conjunctivitis which attacks newly born children.

Ophthalmic (of-thal′-mic). That which pertains to the eye. **Ophthalmic Lens,** a lens to be worn before the eye.

Ophthalmitic (of-thal-mit′-ic). That which applies to inflammatory diseases of the deeper as well as the superficial structures of the eye.

Ophthalmitis (of-thal-mi′-tis). (Gr. ophthalmos = eye + itis.) Inflammation of the eye, more especially the globe with its membranes.

Ophthalmoblennorrhoea (-blen-ur-ree′-ah). (Gr. ophthalmos = eye + blenna = mucus + rhoia = flow.) A flow of mucus from the eye.

Ophthalmocarcinoma (-kahr-si-no′-mah). Cancer of the eye.

Ophthalmocele. (Gr. ophthalmos = eye + kele = hernia.) See Stapyloma.

Ophthalmocopia (-koh′-pee-ah). (Gr. ophthalmos = eye + kopos = fatigue.) Fatigue of the eyes; Asthenopia.

Ophthalmodynia (-din′-e-ah). (Gr. ophthalmos = eye + odyne = pain.) Neuralgic pain of the eye.

Ophthalmography (-mog′-rha-fee). (Gr. ophthalmos = eye + graphe = a description.) A description of the eye.

Ophthalmologist (of-thal-mol′-o-gist). One who practices ophthalmology and has taken the degree, "Doctor of Medicine." An **Oculist.**

Ophthalmology (of-thal-mol′-o-gy). A study of the eye and its diseases.

Ophthalmomacrosis (-ma-kro′-sis). Enlargement of the eyeballs.

Ophthalmomalacia (-ma-la′-sha). (Gr. ophthalmos = eye + malakia = softness.) That condition in which there is abnormal softness of the eyeball.

Ophthalmopathy (-mop′-a-thee). (Gr. ophthalmos = eye + pathos = suffering.) Any disease of the eye.

Ophthalmophthisis (-mof′-thi-sis). (Gr. ophthalmos = eye + phthisis = wasting.) That condition in which there is wasting of the eyeballs.

Ophthalmoplegia (-ple′-je-ah). (Gr. ophthalmos = eye + plege = stroke.) Paralysis of the ocular muscles of the eye. **O. Partial,** a form in which only some of the muscles are paralyzed. **O. Progressive,** a gradual paralysis of all the muscles of both eyes. **O. Total,** when the iris and ciliary body, as well as the external muscles, are paralyzed. **O. Externa,** when the external muscles are paralyzed. **O. Interna,** paralysis of the internal muscles.

Ophthalmoptoma (-mop-to'-mah). Protrusion of the eyeballs.

Ophthalmoptosia (-mop-to'-sia). (Gr. ophthalmos = eye + ptosis = a falling.) Protrusion of the eyeball.

Ophthalmorrhagia (-mor-rha'-gee-ah). (Gr. ophthalmos = eye + rhegnymi = I burst forth.) Hemorrhage from the eye or orbit.

Ophthalmorrhexis (-mor-rex'-is). (Gr. ophthalmos = eye + rhexis = rupture.) The bursting of the eyeball.

Ophthalmoscope (of-thal'-mo-scope). (Gr. ophthalmos = eye + skopeo = I examine.) An instrument for observing the interior of the eye, and thus determining the appearance of the media, the condition of the retina, choroid and optic nerve, and the state of the refraction.

The Ophthalmoscope consists of a round mirror, with a small perforation in the center. The surface of the mirror is usually concave. The more improved ophthalmoscopes have a reversible mirror, one side of which is flat and the other concave. In addition to this there are located on the back of the ophthalmoscope several wheels which contain a great variety of convex and concave lenses. By rotating these wheels the different lenses contained in them can be thrown immediately behind the aperture in the mirror.

There are two methods of examining the eyes with the ophthalmoscope, viz.: the indirect and the direct.

The indirect method is not of much value so far as estimating the refraction of the eye is concerned, but gives a good view of the fundus of the eye, enabling us to examine in minute detail the optic disc and the blood vessels of the retina: also to observe whether any diseased condition exists in the interior of the eyeball.

To perform this method successfully, we seat our patient in the dark room and place a light, either an argand gas burner, an electric light, or any ordinary kerosene student's lamp, at the side and slightly back of the patient's head. If we wish to examine the left eye we place the light on the left side of the patient's head, and if we wish to examine the right eye, on the right side. We place the light just far enough back of the head to avoid illuminating the patient's face. We take our seat in front of the

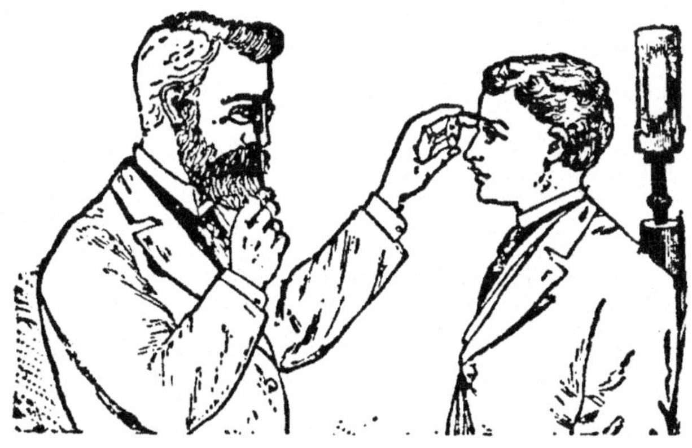

patient and hold our ophthalmoscope at the focal distance of its mirror, reflect the light into the patient's eye, and look through the sight hole in our mirror. The next step is to place a strong convex lens immediately in front of our patient's eye. With our mirror we illuminate the retina, and the rays emanating

disc will neither increase nor decrease in size because the rays which emanate from an emmetropic eye will be parallel, and hence will at all times be brought to a focus exactly at the focal distance of the convex objective glass.

In myopia the image of the optic disc will appear smaller than in emmetropia and much smaller than in hypermetropia. On gradually withdrawing the objective glass the optic disc will increase in size.

In case of astigmatism the optic disc usually looks oval and on withdrawing the glass it will increase or decrease more rapidly in one meridian than in one at right angles to it.

In making the direct ophthalmoscopic examination we also use a concave mirror, and arrange the position of the light similar to that

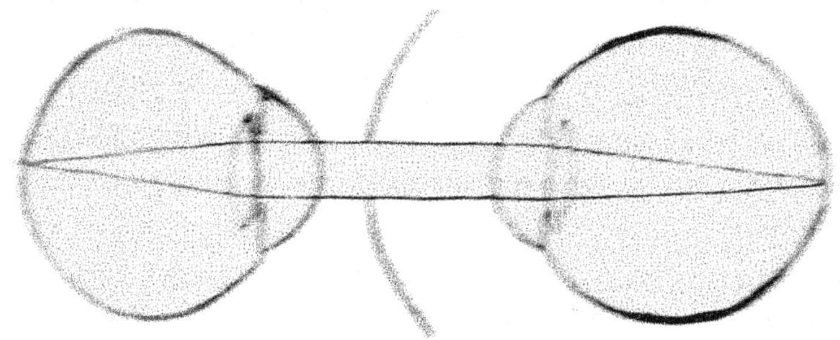

described for making the indirect examination, but in this case we place our own eye very close to that of the patient. We place our ophthalmoscope at a distance of about two inches from the patient's eye, and place our own eye immediately behind the perforation in the mirror. It is necessary in this examination that neither the patient nor the operator shall use any accommodation whatever, and in order that this may be thoroughly accomplished it is well to place the patient under the influence of atropine, assuming that the operator has thorough control of his own accommodation and can suspend it at will.

By reflecting the light into the Emmetropic eye we illuminate the retina; the retina again reflects the light so that it passes through the pupil and out of the eye. The rays from each reflecting point emerge from the eye as parallel rays. Therefore, if the operator's eye is Emmetropic he will, by the aid of these rays, be able to distinctly see the details of the fundus in the back part of his patient's eye, because the light leaves the patient's eye as parallel rays, and the operator's eye, being also emmetropic, is adapted for parallel rays, and hence he has each point on the patient's retina represented by an exact focus on his own retina.

It can readily be seen that if a patient has used any accommodation the rays would leave his eye, not as parallel, but as convergent rays, and therefore the operator would have been unable to distinctly see the details of the fundus. On the other hand, if the patient had been under the influence of atropine and the rays had left the eye as parallel rays striking the observer's emmetropic eye, as parallel rays, they would not focus upon the operator's retina if he had used any accommodation. Hence the necessity of thorough relaxation of accommodation in both the patient and operator.

We will next suppose that the hypermetropic eye is being examined by an operator who is emmetropic. The rays of light will leave the hypermetropic eye as divergent rays and therefore will not focus upon the retina of the observer's emmetropic eye, and hence he will be unable to distinctly see the details of the fundus in the patient's eye. He now rotates the wheel on the back of his ophthalmoscope, throwing different lenses into the aperture of the mirror, until finally he strikes one which enables him to distinctly see his patient's retina. The glass which produces this result will represent the measure of his patient's

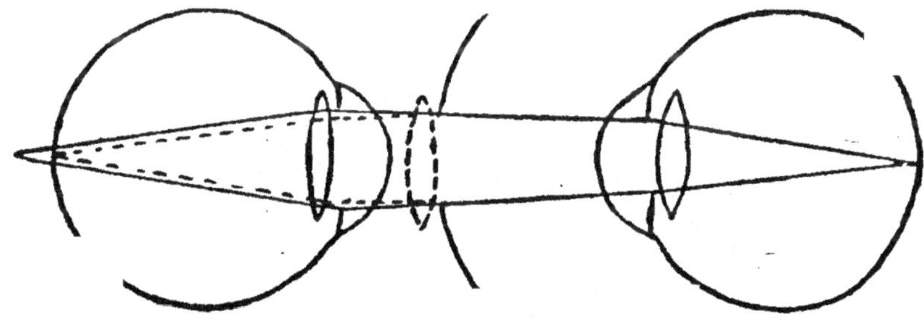

hypermetropia. In order that the operator may distinctly see the patient's retina, he must use a convex glass strong enough to render parallel the divergent rays which are leaving his patient's eye.

In myopia the patient's eye is too long, and the rays of light which leave the myopic eye will therefore leave as convergent rays and focus in front of the observer's retina, so that in this case he will also be unable to distinctly see the fundus of his patient's eye, and as in the case of the hypermetropic eye he rotates the wheel on his ophthalmoscope until he is able to distinctly see the retina of the myopic eye. The glass which accomplishes this result is the measure of the patient's myopia.

In order that he may distinctly see the back

of this eye, a concave glass must be used strong enough to render parallel the convergent rays which are leaving the myopic eye.

It is usually conceded that in astigmatism the ophthalmoscope is of little or no value. It is true that we may be able to see the blood-vessels and the edges of the optic disc clearer in one meridian than in another, and that we may use a glass strong enough to render plain the

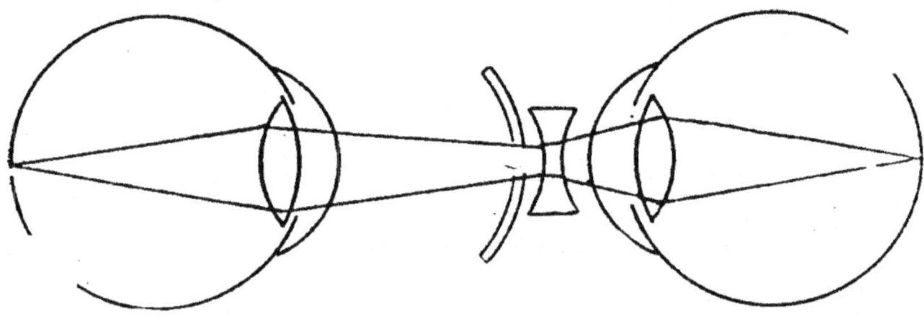

meridian which was at first blurred, and thereby dim the meridian which was first seen plainly. The glass which clears up the meridian which seemed blurred at first would be the measure of the ametropia in the meridian at right angles to it.

It must also be remembered that if the operator is not emmetropic he must either have his own refraction properly corrected by glasses, or he must make deductions or additions, as the case may be, to or from whatever glass clears up the retina in the patient's eye. For instance, if the operator knows himself to be hypermetropic to the extent of one dioptry, and on examining his patient's eye he finds that a plus 3-D. glass is required to enable him to see the details of his patient's fundus, he then knows that his patient has two dioptries of hypermetropia, because one of the three was required to correct his own eye, and likewise if he had been examining a myopic eye and found that a minus 2-D. lens was required to see his patient's retina, he would know that his patient had three dioptries of myopia, because he himself was one dioptry hypermetropic.

Ophthalmoscopy (oph′-thal-mos′-copy). The art of judging healthy or diseased conditions of the eyes by means of the ophthalmoscope. There are two methods of examining the eyes with the ophthalmoscope, viz.: the direct and the indirect.

Ophthalmostat (of-thal′-mo-stat). (Gr. ophthalmos = eye + statos = made to stand.) An eye-speculum.

Ophthalmula (of-thal′-mu-lah). A cicatrix located upon the eye or its appendages.

Optic. (Gr. opsis = vision.) That which pertains to the science of light, and also to the eye itself, together with its functions.

Optic Atrophy. A partial or total loss of sight due to atrophy of the optic nerve.

Optic Axis. A line drawn through the center of the cornea, through the nodal point to the inner side of the macula lutea. In other words, through the center of the eyeball from before back.

Optic Groove. A transverse groove on the upper surface of the sphenoid bone on which the optic nerve rests.

Optic Nueritis. Inflammation of the optic nerve.

Optical. Pertaining to the organ of vision.

Optical Center. A point where the secondary axis crosses the principal axis on line with the thickest part of a plus sphere or the thinnest part of a minus. A ray of light when passing through the optical center of a lens will always emerge parallel to the incident ray or in the same plane. **Optical Center of a Lens.** The center of refraction. It is found by making two parallel radii of curvature, and connecting the points in which they meet the surfaces. The point at which this line cuts the principal axis, is the optical center.

Optical Corrections. Lenses that change the direction of light rays entering the eyes to such an extent that the eyes are adapted to receive and focus them upon the retina, creating artificial emmetropic conditions when ametropic conditions exist.

Optic Commissure (kom′-mis-ur). The linking or joining together of the right and left optic nerve.

Optic Disc. That spot on the retina which marks the entrance of the optic nerves into the eye. It is also called the blind spot, or papilla.

Optic Excavation. The depression in the optic disc.

Optic Nerve. See under nerve.

Optic Papilla (pap-il′-lah). The elevation of the optic-nerve head; also called the optic disc and blind spot.

Optician (op-tish′-an). A person skilled in the manufacture of optical instruments.

Optics. That part of the science of physics which deals with the transmission of light, the laws of refraction, reflection and the phenomena of vision.

Optic Tract. The optic nerve between the visual centers and the optic commissure.

The Optic Nerve, which, in part, is known as the optic tract, is formed by two roots, in the floor of the brain, the external and the internal roots.

The External Root has its origin in three gray matter centers: (1) The optic thalamus; (2) the external geniculate (knee-like) body, and the anterior tubercles of the corpora quandrigemina (four bodies).

The Internal Root arises from two gray matter centers; they are the internal geniculate body and the posterior tubercles of the corpora quadrigemina (four bodies).

These sight centers which give origin to the optic tracts are connected to the cerebral cortex by a number of fibers known as the cortico optic radiating fasciculi and form the most posterior part of the optic thalamus (bed). These later fibers are supposed to be of a sensory nature, and to communicate with the different sections of the brain. The fibers of the external and internal roots come together to form the optic tract, it passes forward along under the posterior of the optic thalamus, crosses the crus cerebri, and again crosses the side of the tuber cinerium to the optic groove on the sphenoid bone where it unites with the optic tract of the other side to form the optic commissure. In the optic commissure the fibers of each optic tract divide, and the outer fibers of each tract are continued into the nerve of the same side, while the central fibers of each tract continue into the optic nerve of the opposite side, crossing each other and passing through the optic foramen to enter the eye.

The dura mater and pia mater which lines the skull passes through the optic foramen (hole into orbit) and forms the optic sheath or covering of the optic nerve until it enters the lamina cribrosa (sieve-like opening of the eye) and spreads out to form the retina.

Optist (op′-tist). A person skilled in optometry.

Optogram (op′-to-gram). (Gr. optos = visible + gramma = a picture.) A faint image stamped on the retina for a brief period.

Optology (op′-tol-o-ge). See **Optometry**.

Optometrist. One who measures the eye's refraction.

Optometry (op-tom′-e-try). (Gr. optikos = belonging to sight + metron = measure.) The science and art of employing the various methods of measuring the optical state of the eye.

O′ra Serra′ta (a serrated border). The anterior limit or edge of the retina. So named from its saw-like appearance.

Orb. A spherical body.

Orbicular. (L. orbiculus = a small disc.) Annular, circular.

Orbicularis Palpebrarum. The circular muscle of the eyelids.

Orbiculus Ciliaris. A zone of about one-sixth of

an inch in width. It is directly continuous with the anterior part of the choroid.

Orbit (or′-bit). (L. orbita = track). The bony socket in which the eyeball is placed. The orbits are conical in shape with their apices extending backward and toward each other while the front or base of the cone is open, leaving the eyeballs to be protected by the eyelids in front.

Anatomy of Orbits. The orbits are two pyramidal cavities 1½ inches wide by 1¾ inches deep, situated at the upper and anterior part of the face, their bases being directly forward and outward and their apices backward and inward, so that the axes of the two if continued backward would meet over the body of the sphenoid bone. The orbit is lined with periosteum, the periorbita. Each orbit is formed of seven bones, the frontal, the sphenoid, ethmoid, superior maxillary, malar, lachrymal and palate; but three of these, the frontal, ethmoid and sphenoid, enter into the formation of both orbits, so that the two cavities are formed of eleven bones only. The orbital opening, or mouth, is called aditus orbitae (Aditus Orbitae—entrance to orbit). At the apex, or back part of the orbit on the nasal side, is a small circular opening known as the optic foramen, which transmits the optic nerve and ophthalmic artery. There are nine openings communicating with each orbit, viz., the optic foramen, the spheno-maxillary fissure, sphenoidal fissure, supraorbital foramen, anterior and posterior ethmoidal foramina, infraorbital canal, malar foramina, and the canal for the nasal duct.

Spheno-maxillary fissure transmits the superior maxillary nerve and its orbital branches, the infraorbital vessels, and the ascending branches from the spheno-palatine or Meckel's ganglion. Sphenoidal fissure transmits the third, the fourth, the three branches of the ophthalmic division of the fifth, the sixth nerve, some filaments from the cavernous plexus of the sympathetic, the orbital branch of the middle meningeal artery, and a branch from the lachrymal artery of the dura mater, and the ophthalmic vein. Supraorbital foramen transmits the supraorbital artery, nerve and vein. Anterior ethmoidal foramen transmits the anterior ethmoidal vessels and nasal nerve. Posterior ethmoidal foramen transmits the posterior ethmoidal vessels. Infraorbital canal opens just below the margin of the orbit. Malar foramina is a passage for nerves and vessels from the orbit.

Orbital (or′-bit-al). Pertaining to the orbit.

Origin (or′-ij-in). (L. origo = beginning.) The more fixed end of a muscle; for instance, the end attached to the bone of the orbit.

Orthochromatic (or-tho-chro-mat′-ic). (Gr. orthos = straight + chroma = color.) A term used by photographers denoting that the colors are normal or correct.

Orthometer (or-thom′-e-ter). (Gr. orthos = straight + metron = measure.) An instrument for finding the exact relative protrusion of the two eyeballs.

Orthoptic (or-thop′-tic). (Gr. orthos = straight + optikos = relating to sight.) Relating to the straightening of a deviating eye by means of exercise.

Orthophoria (or-tho-foh′-ree-ah). (Gr. orthos = straight + phoria = tending.) That condition in which the eyes, or better, the visual axes, are parallel when the extrinsic muscles are in a state of rest. Perfect muscular balance.

Orthoptic (or-thop′-tic). (Gr. orthos = straight + optikos = relating to sight.) Correcting heterophoria, or strabismus, by means of the prism exercise. This is accomplished by placing the base of the prism over the strong muscle, thus causing the weak muscle to contract or draw the eye toward the apex of the prism in order to see.

Orthoscope (or′-tho-scope). (Gr. orthos = straight + skopeo = I view.) An instrument for neutralizing the refraction of the cornea by examining it through water.

Orthoscopic Lenses. A lens with two elements—a sphere and a prism—so arranged that the amount of accommodation and convergence used should exactly correspond.

Orthotropia (or-tho-tro′-piah). (Gr. orthos = straight + trope = turn.) Perfect binocular fixation. With this condition heterophoria may exist and some muscle or muscles are under strain to hold the eyes parallel, yet it is generally accompanied by orthophoria.

Osseous (os′-e-us). Bony. Resembling bone.

O. S. (Oculus Sinister). Left eye.

O. U. (Ocular Unati). Both eyes.

Oxyopia (ox-y-o′-pi-ah). (Gr. oxys = acute + ops = eye.) That condition in which the sight is abnormally acute.

PACHYBLEPHARON (pach-y-blef-ar′-on). (Gr. pachys = thick + blepharon = eyelid.) That condition in which the eyelids have become thickened.

Palpebra (pal′-pe-brah). (L. palpetare = to palpitate.) The eyelid. **Inferior P.,** lower eyelid. **Superior P.,** upper eyelid. **Tertia P.,** third eyelid. See Membrana Nictitians.

Palpebral (pal′-pe-bral). (Gr. palpebra = eyelid.) That which relates to the eyelid.

Palpebral Fissure (pal′-pe-bral). The space between the free margins of the eyelids. The outer angle of fissure is called the external canthus; the inner angle, the internal canthus. The small space between the lids and globe at inner angle is called the lacus lachrymalis.

Palpebritis (pal-pe-bri′-tis). (L. palpebra = eyelid + Gr. itis = inflammation.) An inflammation of the eyelids.

Pannus (pan′-nus). (L. "a piece of cloth.") A web-like patch of grayish membranish tissue usually covering the upper half of the cornea, making it almost opaque. It is usually caused by the rubbing of roughened or granulated lids over the cornea and results in the effort of nature to protect this membrane.

Panophthalmia (pan-of-thal′-mi-ah). An inflammation of the entire eye structure.

Panophthalmitis (pan-off-thal-mi′-tis). (Gr. pas (pan) = all + ophthalmos = eye.) General inflammation of the eyeball.

Pantiscopic. A lens tilted outward at the top.

Pantometer (pan-tom′-e-ter). An instrument for measuring angles and determining perpendiculars.

Papilla (pap-il′-lah). (L. "nipple.") A conic elevation observable at the optic-nerve head. **P. Lachrymalis,** the mound at the inner canthus of the eye pierced by the lachrymal puncta.

Papillitis (pap-il-li′-tis). That condition in which there is an inflammation of the optic disc or papilla.

Papilloretinitis (pap-il-lo-ret-in-i′-tis). Inflammation of the optic disc and retina.

Parablepsis (par-ab-lep′-sis). (Gr. para = beside + blepsis = sight.) False vision.

Paracente′sis. (Gr. para = beside + kentesis = puncture.) Surgical puncture of a cavity.

Paracentesis Cornea (-sen-tee′-sis). Puncture of the cornea.

Parallax (par′-al-lax). An apparent displacement of an object, due to change in the observer's position, or when closing one eye.

Parallel (par′-al-lel). That which pursues the same direction, but in a separate path.

Parallelepiped (par-a-lel-e-pip′-ed). (Gr. parallelos = parallel + epipedon = plane.) A prism whose bases are parallelograms.

Parallelogram (par-a-lel′-o-gram). (Gr. parallelos = parallel + gramma = line.) A quadrilateral whose opposite sides are parallel.

Parallelism (par′-al-lel-ism). State of being parallel. That condition in which the visual axes of both eyes lie in nearly parallel paths.

Paral′ysis. (Gr. "I loosen"—"I relax.") That condition in which there is a loss of power of voluntary motion or of sensation in a part from lesion of nerve substance. **Oculo-motor P.,** where the motor oculi nerve is affected.

Paralysis of Accommodation. That condition in which the function of the branch of the third nerve which supplies the ciliary muscles has been interrupted and the eye cannot accommodate, the ciliary muscles being in a state of rest.

Paralyt′ic. (Gr. affection.) Pertaining to, or affected with, paralysis; a person who is affected with paralysis.

Paresis (par′-es-is). (Gr. "a relaxing.") A slight form of paralysis.

Paropsis (par-op′-sis). (Gr. para = beside + opsis = vision.) That condition in which the vision is disordered, and may be due to either a false impression being made upon the retina or a disordered condition of the mind.

Passive. (L. passivus = to endure.) That which is not active; for instance, a muscle that is in a state of rest.

Pathetic (pa-thet′-ik). That which pertains to the feelings. The pathetic muscle is the superior oblique muscle of the eye, which receives its name from the fact that the patheticus, or fourth pair of cranial nerves, control its movements.

Pathologic. Pertaining to diseased conditions.

Pathological (path-o-log′-i-cal). See Pathology.

Pathology (path-ol′-o-je). (Gr. pathos = suffering + logis discourse.) The science which has for its object the knowledge of disease.

Pediculis Pubis (ped-ik′-u-lus pu′-bis). (L. "crab-louse.") Crab-louse. In very rare cases they will reach the eyelashes and flourish there. The lice cling close to the border of the lid, and look like dirty scabs; the eggs are darker, and may also be mistaken for bits of dirt. The absence of inflammation and the rather peculiar appearance will lead, in doubtful cases, to the use of a magnifying glass, by which the question will be settled at once.

Penumbra (pe-num′-brah). A partial shadow.

Perception. (L. percipĕre = to perceive.) The acquiring of impression through the senses. **Centers of sight p.,** those portions of the brain that are the sources of the optic nerves.

Perceptivity (per-sep-tiv′-it-e). Capacity to receive impressions.

Perfection Bifocal. See Bifocal.

Perichoroidal (-koh-roy′-dul). That which surrounds the choroid membrane.

Pericorneal (per-i-cor′-ne-al). (Gr. peri = around.) That which is situated around the cornea.

Perimeter (per-im′-e-tur). (Gr. peri = around + metron = measure.) An instrument for measuring the field of vision.

Perimetry (pe-rim′-et-re). (Gr. perimetros = circumference.) Measurement of the visual field.

Periocular (per-e-ok′-u-lar). (Gr. peri = around + L. oculus = eye.) That which encircles the eye.

Perioptic. See Periocular.

Perioptometry (per-e-op-tom'-et-re). (Gr. peri = around + optikos = referring to vision + metron = measure.) Measurement of the visual acuity of the retinal periphery.

Periorbita (per-i-or'-bit-a). (Gr. peri = around + L. orbita = orbit.) That which relates to the lining membrane of the orbit.

Periorbital (per-i-or'-bit-al). Around or about the orbit.

Periorbitis (per-e-or'-bi-tis). Inflammation of the lining membrane of the bones of the orbit. Orbital periostitis.

Periosteitis (per-e-os-te-i'-tis). Inflammation of the periosteum.

Periosteum (per-e-os'-te-um). (Gr. peri = around + osteon = bone.) The tough, fibrous membrane investing a bone.

Peripheraphose (per-if-er'-af-oz). The subjective sensation of a dark spot in a patch of light, the cause residing in the eye, optic nerve or outside of optic center in the brain.

Periphacitis (per-i-fa-si'-tis). (Gr. peri = around + phakos = lens + itis.) Inflammation of the crystalline lens of the eye.

Periphacus (per-if-a'-cus). The crystalline lens capsule.

Periphery (per-if'-er-y). (Gr. peri = around + phero = I carry.) Any outward part or surface; for instance, the border of the cornea or crystalline lens.

Periscopic (per-is-cop'-ic). (Gr. peri = around + skopeo = I view.) A lens having a concave and convex surface. Periscopic lenses are also called meniscus lenses; taken from a Greek word meaning a crescent. See Lens.

Peritomy (per-it′-o-me). (Gr. peri = around + tome = incision.) An operation for the treatment of pannus, by removing a strip of the conjunctiva around the cornea.

Perivascular (per-i-vas′-ku-lar). (Gr. peri = around + L. vasculum = vessel.) Surrounding a vessel.

Perivasculi′tis. (Gr. peri = around + L. vasculum = vessel + itis.) Inflammation of the sheath of a vessel. This is an increase or a hyperplasia of the connective tissue about the vessels, principally, and usually, the arteries.

Perspicilium (per-spic-il′-i-um). An apparatus to enable an individual to see minute bodies, or which will improve the eyesight.

Pescorvi′nus. That which is commonly known as crow's foot; or wrinkles at the outer corner of the eye.

Petit. François Pourfour du Petit, French surgeon and anatomist, 1664-1741.

Petit's Canal. The space between the suspensory ligaments in which the edge of the crystalline lens with its capsule is inserted.

Phaco (fak′-o). Prefix meaning of, or pertaining to, a lens, especially the crystalline lens.

Phacitis (fas-i′-tis). (Gr. phakos = lens + itis.) Inflammation of the crystalline lens.

Phacomalacia (fak-o-mal-a′-she-ah). (Gr. phakos = lens + malakia = softness.) A soft cataract.

Phacometer (fa-com′-e-ter). (Gr. phakos = lens + metron = measure.) An instrument for measuring the curvature of lenses, and so determining their refractive power; if they are cylindrical, will locate their axes.

Phacosclerosis (fa-ko-scle-ro′-sis). (Gr. phakos = lens + sklerosis = hardening.) Hardening of the crystalline lens.

Phacoscope (fa′-ko-scope). (Gr. phakos = lens + skopeo = I view.) An instrument used for viewing the accommodative changes of the crystalline lens.

Phakitis (fa-ki′-tis). (Gr. phakos = lens + itis.) Inflammation of the lens. A supposition exists that the crystalline lens may become inflamed.

Phantasma (fan′-tas-mah). (Gr. phantasma = an appearance.) A disease of the eye in which imaginary objects are seen.

Phengophobia (fen-go-fo′-bi-ah). (Gr. phengos — daylight + phobos = fear.) See Photophobia.

Phimosis (fi-mo′-sis) **Constriction.** (Gr. "to muzzle.") Abnormal smallness (as of the palpebral fissure).

Pho′rotone. (Gr. phora = motion + tonos tension.) An instrument for exercising the muscles of the eye.

Phosgenic (fos-jen′-ik). (Gr. phos = light + gennao = to produce.) Light producing.

Phlyctenula (flik-ten′-u-lah). (Gr. "blister.") A small vesicle or blister.

Phlysis (fly′-sis). A corneal ulcer.

Phoria. (Gr. "a tending.")

Phorometer. An instrument for determining the insufficiencies of the external ocular muscles.

Phorometroscope (phor-o-met′-ro-scope). An instrument for determining the amount, correction and treatment of muscular asthenopia by gymnastic exercise of the extrinsic muscles.

Phoroscope. An instrument in the form of a head-rest, with a clamp attached so that it may be fastened to a table, and is used as a fixed trial frame.

Phose (foz). (Gr. phos = light.) A subjective sensation of light or color.

Phosphenes (fos′-feenz). (Gr. phos = light + phaino = I show.) A luminous sensation caused by pressing on the eyeball.

Phosphorescence (fos-fo-res′-ens). (Gr. phos = light + phoros = bearer.) The quality of becoming luminous in the dark without sensible heat.

Photalgia (fo-tal′-je-ah). (Gr. phos = light + algos = pain.) Pain in the eye arising from too much light.

Photochromatic (fo-tō-chro-mat′-ic). (Gr. phos = light + chroma = color.) That which pertains to various colored lights.

Photodysphoria. See photophobia.

Photogenic. See Phosgenic.

Photology (fo-tol′-o-gy). The science of light.

Photometer. (-tom′-e-ter). (Gr. phos = light + metron = measure.) An instrument for testing the light sense.

Photonosus (fo-ton′-o-sus). (Gr. phos = light + nosos = disease.) Any disease of the eye which arises from exposure to the glare of light.

Photophobia (fo-to-fo′-bi-ah). (Gr. phos = light + phobos = fear.) Intolerance of light.

Photopsia (fo-top′-si-ah). (Gr. phos = light + opsis = vision.) That condition in which one sees flashes of light. It is caused either by pressure on the eyeballs or by disease of the brain, optic nerve, or retina.

Photoptometer (fo-top-tom′-e-ter). (Gr. phos = light + optos = visible + metron = measure.) A device for measuring sensitiveness to light by showing the smallest amount of light that will allow an object to become visible.

Phthisis Bulbi (tis′-sis). (Gr. a wasting.) Shrinkage of the eyeball.

Physiolog′ical. See physiology.

Physiology (fiz-e-ol′-o-je). (Gr. physis = nature + logia = discourse.) That department of natural science which treats of the organs of the body and their functions.

Physostigmine (fi-so-stig′-min). The same as eserin.

Pia Mater (L. "tender, affectionate mother"). The innermost membrane of the brain and spinal cord, optic sheath, and capsule of Tenon.

Pigment (L. pingere = to paint.) The coloring matter in the choroid coat; the iris, etc.

Pilosebaceous (pi′-lo-se-ba′-ce-ous). (L. pilus = hair + sebum = suet.) Relating to the hair follicles and sebaceous glands.

Pinguecula (ping-gwek′-yu-lah). (L. pinguis = fat.) A small, yellowish elevation, situated in the conjunctiva near the margin of the cornea. Found in old age.

Pinhole Disc. An opaque disc with a pinhole in the center, found in the trial test case. It is placed in the trial frame quite close to the eye under examination. This perforation gives passage to a small pencil of light which passes through the center of the refracting media of the eye. If the patient can see better through the pinhole, the refracting system is at fault, and vision can be improved by glasses. If, on the contrary, vision is not improved, then we suspect a defect in the sensibility of the retina or the transparency of the media of the eye.

Pink Eye. A catarrhal conjunctivitis. The eyeball is of a pink or reddish color. It is a contagious disease which occurs among cattle and horses as well as in man.

Pladaro'sis (Gr. pladaros = flaccid + oma = tumor). That condition where there is a soft tumor on the eyelid.

Plane (L. planus = flat). When applied to glass, a flat surface is meant. A plano-concave lens is a lens having one side concave while the other side is flat. A plane disc, or a plano, is an accessory found in the trial case which has two surfaces, both of which are plane.

Plastic (plas'-tik). (Gr. plastikos = form.) Tending to build up tissues.

Plexus (plex'-us). (L. plectere = "to weave.") A network or interjoining of nerves or vessels.

Plica (L. plicare = to fold). A fold. Applied to a disease in which the hairs become tangled and glued together.

Plica Semilunaris (ply'-kah). A fold of conjunctiva near inner canthus of the eye.

Point (L. punctum). The far point or **punctum remotum** is the farthest point at which the eye can see clearly and distinctly with the accommodation at rest. The near point or **punctum proximum** is the nearest point at which the eye can see clearly with all of its accommodation in use. **P. of Reversal.** In Retinoscopy the term is used to designate the point between an erect and an inverted image, where the change from one to the other occurs. Where convergent rays change to divergent rays. The myopic far point in retinoscopy is where the movement of the reflex appears neutralized. In other words, it is that point on one side of which the shadow movement is different than on the other. For instance, at any position nearer the eye than the point of reversal the shadow will move against the mirror, and at any position farther from the eye the shadow will move with the mirror. This refers to the concave retinoscope. With the flat mirror the movement would be directly opposite. **P. of Fixation.** The point for which accommodation of the eye is adjusted.

Polarimeter (L. polaris = polar + Gr. metron = measure). An instrument for measuring the rotation of polarized light.

Polariscope (L. polaris = polar + skopeo = I examine). An instrument used in showing the phenomena of the polarization of light.

Polarization. The production of a condition in light by virtue of which all its vibrations take place in one plane, or in circles and ellipses.

Pole (L. polus = pole). The summit of a spherical surface.

Polychromatic (pol-y-chro-mat′-ic). (Gr. polys = many + chroma = color.) Possessing many colors.

Polycoria (pol-e-ko′-re-ah). (Gr. polys = many + kore = pupil.) The presence of more than one pupil.

Polyopia (pol-e-o′-pe-ah). (Gr. polys = many + ops = eye.) Multiple vision.

Polyoptrum (pol′-y-op′-trum). (Gr. polys = many + optos = seen.) A glass through which objects appear multiplied but reduced in size.

Pop-Eyed. A large protruding condition of the eyes.

Pore (Gr. "passage"). The superficial opening of a vessel; one of the small openings existing in all bodies.

Po′rus Opticus (L. porus = pore + Gr. opticus = optic). The opening through the lamina cribrosa through which the arteria centralis retina and veins pass.

Positive. That condition which is real and absolute. The positive surface of a periscopic lens is tne convex surface.

Posterior (L. post = after). Behind; back.

Postocular (L. post = behind + oculus = eye). Posterior to the eyeball.

Postocular Neuritis (L. post = behind + oculus = eye + Gr. neuron = nerve + itis). Inflammation of part of optic nerve behind the eyeball.

Presbyopia (pres-by-o′-pi-ah). (Gr. presbys = old + ops = eye.) When as the result of age the power of accommodation has diminished to such an extent that the eye (corrected for distance, if ametropic) cannot produce sufficient accommodation for the reading distance, the condition is called presbyopia. The average age when this state of affairs is present is 45, and as age advances the accommodation gradually diminishes and the presbyopia correspondingly increases. The amount of presbyopia is represented by the difference between the number of dioptries of comfortable accommodation present and three dioptries, which must be made good by plus spheres. The presbyope sees well at a distance, providing there is no error of refraction, but has difficulty in maintaining good vision for near work, and the eyes become tired after reading, especially at night. He has trouble in seeing small objects because he has to hold them far away, and consequently gets a smaller visual angle. Before correcting presbyopia it is necessary to test the patient's distant vision and correct any error of refraction. Then place the reading chart in his hand; if he cannot read with comfort at the distance he wishes to hold it, add plus spheres of even amount in front of his correction until you find the weakest that will allow comfort in reading. The distance for which the presbyope requires glasses will also vary much according to his or her occupation; ordinarily it is thirteen inches.

This gradual failure of accommodation is due to hardening of the crystalline lens, loss of power of the ciliary muscle, or both.

Prescription (pre-scrip′-shun). (L. prae = before + scribere = to write.) The formula for the lenses required by a patient, which are designated by technical characters placed on blanks arranged for this purpose.

Principal Focus. The focus of parallel rays of light on the principal axis after being reflected or refracted.

Principal Meridians. The meridians of greatest and least curvature.

Principal Planes. Straight lines which pass through the principal points, perpendicular to the principal axis.

Prism. When applied to optics, is a wedge-shaped, transparent body of glass having two plane sides, employed for the purpose of bending rays of light. A prism is not a lens, and a ray of light is always bent towards its base. It is used in making tests for muscular insufficiencies, and sometimes prescribed for constant wear in cases of heterophoria.

Prisms are numbered by the angle which their surfaces incline toward each other; for instance, four 90° prisms with their bases and apices placed together would form a circle. The bending power the prism possesses depends upon the difference of density of the glass itself and the medium which it is in. The ordinary prism is made of crown glass and deviates a ray of light about half of its own value; that is, a 4° prism would deviate a ray of light 2°.

Dennett in his method of measuring prisms calls his unit the centrad, which is the hundredth part of a radian, a radian being the angle subtended at the center of a circle by an arc, which is equal in length to the radian.

Prentice Method is the prism dioptry, which

is any prism that has the power to deflect a ray of light 1 cm. for each meter of distance.

These three methods of numbering prisms differ very little for low degrees in ophthalmology.

Rotating Prisms. If two prisms of equal strength be placed with the base of one over the apex of the other, they neutralize each other, and if we rotate them in opposite directions we obtain the effect of any prismatic degree up to their combined values. A prism forms no image and has no focus, and when looked through, the eye turns toward the apex. **Deaton P.**, a prism attached to a microscope to give the oblique illumination for observing very fine markings. **Lateral P.**, an equal-sided, total reflecting prism for illuminating a microscopic field.

Prism-dicptry, n. In **Optics**, a standard deflection of a beam of parallel rays of light produced by a prism. It is equal to 1 cm. on a tangent plane placed at a distance of 1 m. **behind** the prism. To practically measure this deflection while looking through a prism or lens, and consequently upon a tangent plane placed **in front** of the prism, it is necessary to multiply these dimensions by six, in order to insure parallel incidence of the rays constituting the beam of light. The prism-dioptry establishes a definite relation between the refractive powers of prisms and lenses, since "the prism-dioptries in decentered lenses are in direct proportion to their refractive powers and decentration (see Decentration). The prism-dioptry also bears a unique relation to the meter angle (see M. Ang.). The sign used to designate the prism-dioptry is a triangle. Thus the unit, 1Δ of the dioptral system is distinguished from 1° of the old degree system. Since 1895 American lens manufacturers have adopted the prism-dioptry as the standard unit of prismatic power.

Prismatic (pris-mat′-ic). That which has the shape or effect of a prism. When a lens is decentered it will produce a prismatic effect.

Prismoid (priz′-moid). A body that resembles a prism in form.

Prisoptometer (pris-op-tom′-et-er). (Gr. prisma = prism + optos = seen + metron = measure.) An instrument used for testing the refraction of the eye by means of a revolving prism.

Probe. A long, slender instrument for exploring wounds. **Lacrimal P.** is a probe designed for use on the tear passages.

Problem (prob′-lem). (Gr. problema = a question proposed for solution.)

Product (prod′-ukt). (L. pro = forward + ducere = to lead.) The result from multiplying one number by another.

Progressive Myopia. Myopia that is gradually on the increase.

Prophthalmos (prof-thal′-mos). (L. pro = forward + ophthalmos = eye.) A bulging forward or undue prominence of the eyeball.

Proportion (pro-por′-shun). (L. pro = before + portio = share.) A proportion is an expression of equality of ratios.

Proptosis (prop-to′-sis). (Gr. pro = forward + ptosis = falling.) A falling down or sinking of a part.

Prosthesis (Gr. in addition to + to put). The addition of an artificial part to supply that which is wanting.

Prosthesis Ocularis (pro-the′-sis). The insertion of an artificial eye.

Protractor Scale (L. protrahere = to draw forth). A device for indicating the location of the axis of a cylinder lens.

Pseudoblepsis (seu-do-blep′-sis). (Gr. pseudes = false + blepsis = vision.) That condition in which objects look different from what they really are.

Pseudoglioma (seu′-do-gly-oh′-mah). A circumscribed collection of pus in the vitreous.

Psorophthalmia (soh-rof-thal′-mee-ah). (Gr. psoros = scabby + ophthalmos = eye.) That inflammatory condition of the eye which is accompanied with itchy ulcerations.

Pterygium (ter-yg′-i-um). (Gr. pteryx = wing.) A thickening or growth of the conjunctiva. having the appearance of a fly's wing, usually on the nasal side of the eye, extending out toward the cornea. It can be removed by operation, and should be as soon as it reaches the cornea, otherwise it will grow over it and impair vision.

Pterygoid (ter′-ig-oid). (Gr. pteryx = wing + eidos = resemblance.) Wing-shaped.

Ptilosis (ti-lo'-sis). That condition where there is a falling out or loss of the eyelashes.

Ptosis (to'-sis). (Gr. ptosis = a falling.) A drooping of the upper eyelid. This condition is caused by paralysis of that branch of the third or motor oculi nerve which supplies the levator palpebra muscle. It may also be caused by the thickening of the upper lid.

Puncta (punc'-tah). (L. punctum = a point.) A small prominence or point. See Puncta Lacrimalia.

Puncta Lacrimalia (punc'-ta lak-ri-mal'-i-ah.) (L. punctum = point + lachryma = tear.) Two small openings near the nasal end of the surface of each eyelid, through which the tear passes into the lachrymal canal.

Punctum (punc'-tum). A fixed point. See Punctum Remotum and Punctum Proximum.

Punctum Remotum. See Far Point.

Punctum Proximum. See Near Point.

Pupil (pu-pil). (From L. pupa, a babe; so called from the small image seen in the eye.) The circular opening in the iris through which all the rays of light pass that have to form an image of the object on the retina. This aperture is dilated and contracted so as to regulate the amount of light entering the eye. The pupil of man is round, and by it the anterior and posterior chambers of the eye communicate with each other. A contracted pupil (myosis) indicates inflammation of the brain; a sensitive retina, faculative hypermetropia, effect of opium or other drugs. A dilated pupil (mydriasis) indicates effect of atropine or other drugs, myopia, amblyopia, absolute hypermetropia, glaucoma, or paralysis of third nerve.

Anisocoria. Unequal pupils.

Corectopia. Displacement of the pupil.

Coreclisis. Closure of the pupil by a membrane, which causes loss of visual acuity.

Coremorphosis. The operation for artificial pupil.

The shape of the pupil is changed by synechiae, coloboma, iridodialysis, ruptures of the sphincter muscle. The pupil appears black when no light returns through it to the eye of the observer. It is more dilated in youth than in the aged.

Pupillary (pu'-pil-lar-ry). Pertaining to the pupil.

Pupillometer (pu-pil-om'-et-er). (L. pupilla = pupil + Gr. metron = measure.) An instrument for measuring the diameter of the pupil.

Pupilloscopy (L. pupilla = pupil + skopeo = I view). See Retinoscopy.

Pupillostatometer (pu-pil-o-stat-om'-et-er) (L. pupilla = pupil + Gr. statos = placed + metron = measure.) An instrument to measure the distance between pupils.

Pyrkinges Images. The images seen on surface of cornea and lens. See Catoptric Test.

Pyrometer (py-rom'-e-ter). (Gr. pyr = fire + metron = measure.) An instrument for measuring high degrees of heat.

QUADRILATERAL (kwod-ri-lat′-e-ral). (L. quatuor = four + latus = a side.) A four-sided plane figure.

Quantity (kwon′-ti-ti). (L. quantus = how much.) Any amount, in measure or extent.

Quiz (L. quaesitio = inquisition). Instruction by questions and answers. **Q. Class,** a body of students forming a class for the purpose of being questioned by a teacher. (See last few pages.)

Quotient (kwo′-shent). (L. quotiens = how many times.) The number resulting from the division of one number by another.

RACEMOSE (ras′-e-mos). (L. racemus = a bunch of grapes.) Bunched; clustered; as in staphyloma, where the bulging occurs in several places.

Radiad (ra′-de-ad). Towards the radial side.

Radial (ra′-de-al). Of or pertaining to the radius.

Radian. An arc of a circle which is equal to the radius, or the angle measured by such an arc.

Radiant (ra′-de-ant). (L. radiare = to shine.) Diverging, as rays from a center.

Radiation (ra-di-a′-shun). Where rays of light appear to be thrown off from a common center.

Radius (L. "spoke"). The half of the diameter of a circle.

Range of Accommodation. The distance of a patient's vision, or the range between the near point and the far point of vision.

Range of Vision. The distance between the near and far point.

Ratio (ra′-sho). (L., from reri, ratus = to reckon.) The relation which one quantity or magnitude has to another of the same kind. It is expressed by the quotient itself, making ratio equivalent to a number. The term ratio is also sometimes applied to the difference of two quantities as well as to their quotient, in which case the former is called arithmetical ratio, the latter geometrical ratio. Ratio of a geometrical progression, the constant quantity by which each term is multiplied to produce the one succeeding.

Ray. The smallest imaginary line of light.

Reciprocal Numbers (L. reciprocus = alternating). Two numbers which multiplied together make unity.

Rectangle (rek′-tang-gl). (L. rectus = right + angulus = angle.) A quadrilateral all of whose angles are right angles.

Rectus (L. straight). Applied especially to certain straight muscles.

Red-Blindness. That condition in which a person is unable to distinguish red

Reduction (re-duk′-shun). (L. re = back + ducere = to bring.) Changing the denomination of numbers. **Reduction Ascending,** changing to a higher denomination, as from 144 inches to 12 feet. **Reduction Descending,** changing to a lower denomination.

Reflection (re-flec′-shun). (L. re = back + flectere = to bend.) Throwing back light. Reflection from a plane surface gives an erect image, and the angle of reflection is always equal to the angle of incidence. The image is formed at a distance behind the reflecting surface equal to the (not so with curved mirrors) distance of the object in front of it, and is called a virtual image.

Reflection by a concave mirror. Parallel rays falling on a concave surface are reflected as convergent rays which meet at a point called the principal focus, which is equal to half the radius. The distance of the focus from the mirror is called its focal length.

Reflection from a convex surface. Parallel rays falling on a convex surface diverge and never meet. No matter what the position of the object before a convex mirror, the image is always virtual, erect, and smaller than the object.

Reflector (re-flec′-tor). A device for reflecting light.

Refracting Media (see Media). **R. System.** A lens, or combination of lenses, for the creation of optical images.

Refraction (re-frac′-shun). (L. re = back + frangere = to break.) The bending of a ray of light in passing obliquely from one medium to another of different density. This bending is caused by one side of the ray having its speed increased or decreased according to the density of the second medium. Refraction never takes place in any one medium, but between the media. Light in passing from a rarer to a denser medium is bent toward the perpendicular, and from a denser to a rarer is bent away from the perpendicular. **Double R.**, refraction in which the incident ray is divided into two refracted rays. **Static R.**, refraction of the eye

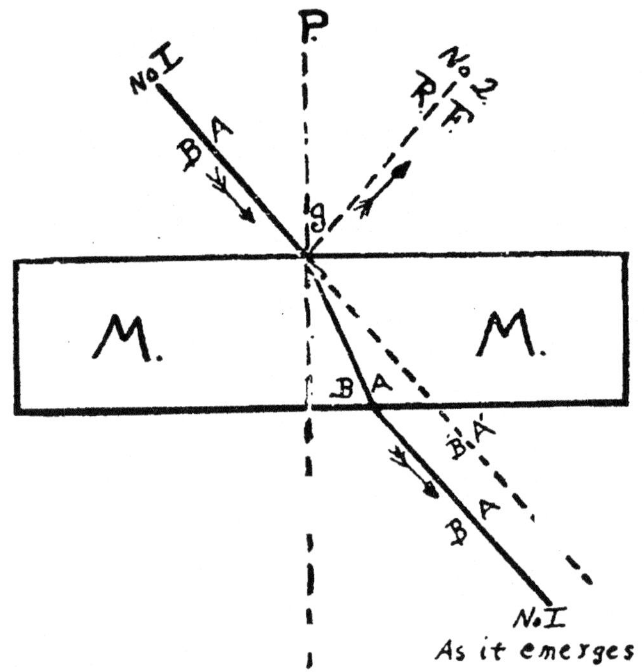

REFRACTION BY PLANE SURFACE.

No. 1 ray of light is called the incident before entering the second medium. A ray passing from a rarer to a denser medium is refracted towards the perpendicular, as shown in the above cut. The ray BA is refracted on striking the glass MM, and again refracted on emerging. In passing from a denser to a rarer medium, the ray is refracted from the perpendicular. P represents a ray falling perpendicular to the surface separating the two media. It continues its course without undergoing any refraction.

No. 2 represents the reflected ray. The angle formed by the incident ray with the perpendicular is always equal to the angle of reflection.

The dotted line marked BA' represents the course the No. 1 ray would have taken had it not been refracted.

The side of the incident ray marked B will be found at R in the reflected ray, and A at F.

at rest. **Dynamic R.**, refraction of the eye, plus that secured by accommodation.

Absolute Index of Refraction is that which is

found when light passes from a vacuum into a given medium.

Refractionist (re-frac′-tion-ist). One who is skilled in correcting errors of refraction of the eye.

Refractive (re-frac′-tive). Pertaining to refraction.

Refractometer (re-frak-tom′-e-ter). An instrument for measuring refraction.

Regular (reg′-u-lar). (L. regula = a rule.) According to rule; normal.

Relative Index of Refraction is that which is found when light passes from atmospheric air into another medium.

Relax (L. re = back + laxare = to loosen). To loosen, to slacken.

Relaxa′tion. A lessening of tension.

Remedy (L. re = again + mederi = to heal). Anything acting as a cure for, or the relief from, unhealthy conditions.

Reposition (re-po-zi′-shun). (L. repositus = to lay up.) The act of putting back in a normal position.

Retina (ret′-in-a). (L. rete = a net.) On the inner surface of the choroid, and closely in contact with it, we find the internal or third and most important of the ocular tunics, the retina; to which, indeed, the other two are merely protective or containing membranes. The retina is the immediate continuation of the optic nerve, which extends from the brain to the eyeball, perforates the sclerotic and choroid, and immediately spreads out into a thin lamina over the surface of the latter, and is attached at two points only —at the entrance of the optic nerve and at its most anterior border, the ora serrata. The point of entrance of the optic nerve, which is known as the optic disc, is nearly on the horizontal meridian of the globe, and about one-tenth of an inch to the nasal side of the posterior pole, so that it is the left eye which is represented in

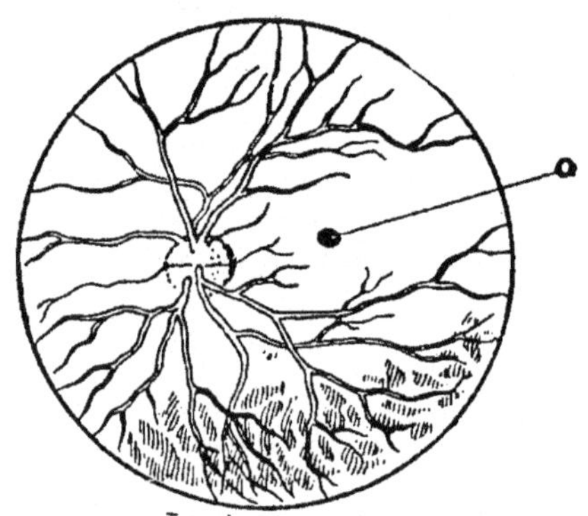

RETINA.
a—macula lutea, the most sensitive part of the retina.

the cut under Anatomy. The function of the retina is to receive the pictures which are formed within the eye by means of the waves of light reflected from objects, and, through the medium of the optic nerve, to transmit the resulting visual impressions to the brain.

Just as the sense of touch is not diffused uniformly over the surface of the body, but is more acute in some parts—for instance, the finger tips—than in others, so also the retina is not equally sensitive to the luminous impressions over its whole surface, but in the highest degree a little to the temple side of the posterior pole, in a part known as the macula lutea, or yellow spot, which may be considered the real center of the retina, yet it is to one side. From this spot the sensitiveness gradually diminishes to its most anterior edge. The retina does not extend as far forward as the choroid, but terminates a little in front of the equator, at the posterior border of the ciliary body, in a saw-like margin, the rough edge of which is known as the ora serrata.

Structures of the Retina According to Gray. From within outward, the layers of the retina are named as follows:

1. Membrana limitans interna.
2. Fibrous layer, consisting of nerve fibers.
3. Vesicular layer, consisting of nerve cells.
4. Inner molecular, or granular layer.
5. Inner nuclear layer.
6. Outer molecular, or granular layer.
7. Outer nuclear layer.
8. Membrana limitans externa.
9. Layer of rods and cones, or Jacob's Membrane.
10. Pigmentary layer.

Retinal Reflex. A term used in retinoscopy to designate the light reflected from the retina and creating the light in the pupil.

Retinitis (ret-in-i'-tis). (L. retina + Gr. itis = inflammation.) Inflammation of the retina. It is characterized first of all by a diffused cloudiness of the organ. The cloudiness varies very greatly in intensity, although in general it is greatest in the vicinity of the optic disc, because here the retina is thickest. Consequently, the outlines of the optic disc become indistinct and the vessels in the retina hazy. The function of the retina is impaired in proportion to the intensity and extent of the inflammation. In the lightest cases vision may be normal, so that the patients complain simply of the presence of a light-colored cloud before their eyes. But for the most part vision is very considerably reduced, both because of the changes in the retina itself and because of the accompanying opacities in the vitreous. The course of retinitis is always rather sluggish. It is only in the lightest cases that the inflammation abates completely within a few weeks, and then the visual acuity may once more become perfectly normal. But for the most part it takes several months for all the inflammatory symptoms to disappear from the retina, while the sight remains permanently impaired. Severe and, more particularly, recurrent inflammations of the retina lead to atrophy of it, pigmentation frequently occurring at the same time (through migration of pigment from the pigment-epithelium). When atrophy of the retina has once made its appearance, the sight is always destroyed, either completely or all except a small remnant, and its restoration is no longer possible.

Retinoscope (ret′-in-o-scope). An instrument with which an objective examination of the dioptric state or condition of the eyes may be measured. (Made in plane and concave.)

The concave can be combined with a strong plus lens, about 20-D., and used as an ophthalmoscope. There is also a difference in the movement of the shadow in retinoscopy. In working with the plane mirror between 53 and 60 inches, the movement is against in myopia of .75 or more, while in hypermetropia, emmetropia, or less than .75 of myopia the shadow moves with the mirror. With the concave it is just the reverse; the shadow in hypermetropia, emmetropia, and a small amount of myopia will go against the mirror. In more than .75 of myopia, the movement will be with the mirror.

It makes no difference which you use, the findings will be the same, and you deduct from plus and add to minus findings the same amounts; it depends on the distance you are

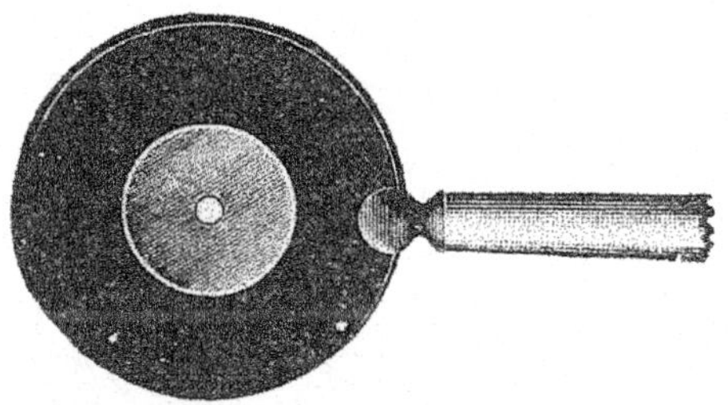

Hand Retinoscope.

sitting from the patient. When sitting at a little over 40 inches, you subtract one dioptry from all plus findings, and add minus .75 to all minus findings. When working between 53 and 60 inches, you subtract .75 from all plus findings and add minus .50 to all minus findings.

Retinoscopy (ret-in-os′-co-py). (L. retina + scopeo = I examine.) "Skiametry." Retinoscopy, or the Shadow Test, is one of the methods of estimating the refraction of the eye. We examine the movements of the shadow when the fundus is illuminated by light thrown into the eye from a mirror.

With the Concave Mirror. The patient is seated in a dark room, with the light placed a little above the head, and far enough back so that it will throw no direct rays upon his face. It is best to use a shade around the light to prevent it from illuminating the walls of the room, having a hole an inch in diameter in the front and about the center of the flame. We will now begin with the examination.

In examining the right eye have the patient look across your right shoulder, and in examining the left have him look over your left shoulder.

Once in a position to begin the test, we reflect

the light from the mirror across several meridians of the patient's eye, at the same time watching the reddish fundus reflex in the pupil. If the movement of the fundus reflex is against that of the mirror in any one meridian, put a plus sphere before the eye, and continue to increase its strength until you find the **weakest** lens that will reverse the last meridian, whose movement was against that of the mirror.

If there is no astigmatism the reflex will be equally bright in all its parts like a small full moon. On the other hand, if there is any astigmatism the shadow will have a band or ribbon-like appearance, the sphere being a finding for the meridian of the band.

We next proceed to correct the meridian at right angles to the band, using a minus cylinder (on account of the movement being with that of the mirror) with its axis over the band, and continue to increase its strength until we find the **weakest** that will open up the band until the reflex is round in appearance.

From the sphere now before the eye we deduct the power of a lens that will focus at the distance the mirror was held from the patient's eye. What is left of the sphere combined with the cylinder is the patient's Rx.

Myopia: On the other hand, if the shadow moves **with** in all meridians, put on a weak minus sphere and increasing its strength until we find the weakest that will about reverse the shadow in any one meridian. If there is no astigmatism the shadow will move the same in all meridians and have the appearance of a full moon. If there is any astigmatism the reddish fundus reflex will have a ribbon-like appearance (the narrower the band the higher the amount), the sphere always being the finding for the meridian in which the band is seen. The meridian at right angles to the band is still moving with the movement of the mirror and a minus cylinder with its axis over the band is used to reverse it while moving the mirror across the band.

To the sphere already in the trial frame we add a minus sphere that will focus at the distance we are holding the mirror from the eye of a patient.

In writing the Rx, put down what is left of the sphere after deducting or adding for the working distance, combining with it the full power of the cylinder, placing its axis at the degree the scratch on the lens points to on the trial frame.

If the above directions are followed plus cylinders will never be used.

The larger the error the slower the movement will be. Large errors are easier than small ones to detect and correct with the retinoscope.

Prove up your retinoscopic test subjectively. If more plus can be added or less minus given without interfering with the vision, make the change.

In using the plane mirror follow above directions, but use plus when the shadow moves **with** and minus if **against.**

When the light that is coming from the patient's eye (after reflection by the retina) focuses in front of the operator the shadow will move with the movement of the concave mirror. On the other hand, if these rays of light pass the operator without focusing, the movement will be against that of the mirror

The shadow moves against in Hypermetropia, Emmetropia and small amounts of Myopia. If the operator is working at 41″ the eye must have one dioptre or more of Myopia for the shadow to move with the movement of the mirror.

On page 223 are two cuts showing the fundus reflex in the pupil. The reflex shows white instead of red as it really is but they will answer the purpose; the first is round on the edge like

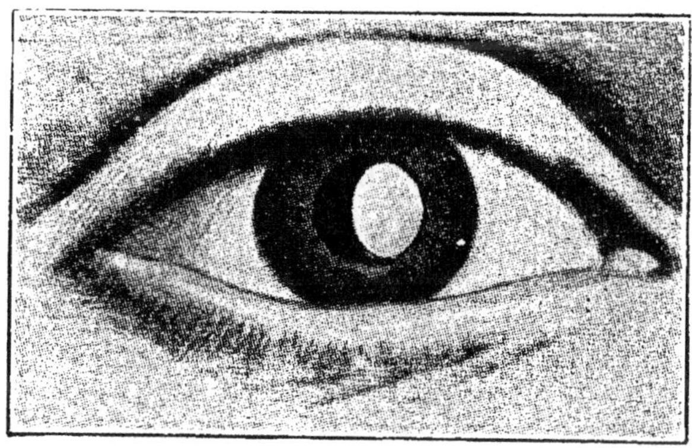

a full moon, indicating no sign of astigmatism, while the second is ribbon or band like, showing there must be astigmatism between the 90th and 180th meridians in this case.

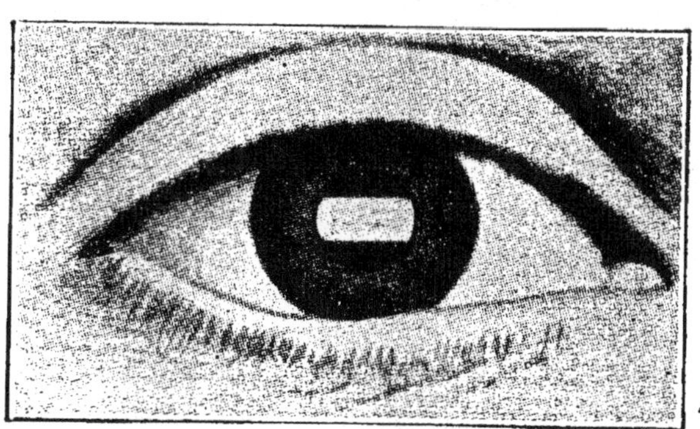

Retractor (re-trac'-tor). An instrument used for drawing and holding the parts away while undergoing an operation, or for any other purpose.

Retrobulbar (re-tro-bul'-bar). (L. retro = behind + bulbus = bulb.) That which is situated or occurring behind the eyeball.

Retrobulbar-Neuritis. Inflammation of the optic nerve behind the globe of the eye.

Reversal Point. This term is used in retinoscopy to describe the change of movement of the shadow. For instance if the rays of light which are coming from the patient's eye focus behind the operator the shadow will always move against the movement of the concave mirror, if they are made to converge by the aid of a plus lens so as to focus just in front of the operator, the movement will be with that of the mirror, making it the point of reversal.

Rheum (rume). (Gr. rheuma = a flux.) A watery discharge from the eyes.

Rheumatic Iritis. Iritis caused by rheumatism.

Rhodopsin (rho-dop′-sin). (Gr. rhodon = rose + ops = eye.) Visual purple; pigment of outer segment of retinal rods.

Rhytidosis (rit-id-o′-sis.) (Gr. rhytidosis = a wrinkling.) A wrinkling, as of the cornea.

Riolanis Muscle. An involuntary muscle used in closing the eye. It reinforces the orbiculars and brings the margin of the lids closer together.

Rod. Relates to the retina. It is one of the cell elements of which Jacob's Membrane is composed. These minute bodies are cylindrical in form, hence the term, rod (rods and cones).

Rodent Ulcer. A destructive ulcer beginning at the margin of the cornea. It is confined to the surface, not going deeper than Bowman's Membrane. First described by Dr. Mooren.

Roentgen Rays. A form of X rays used in determining the presence and also the exact position of foreign bodies in the eyeball or orbit.

Romberg's Symptoms (Moritz Heinrich Romberg, Berlin physician, 1795, 1873). Difficulty in standing when the eyes are shut: a sign of locomotor ataxia.

Rota'tion (L. rota = a wheel.) Process of turning around an axis. **R. of the Mirror.** A term used in retinoscopy to indicate the movement of the mirror to create a movement of the light area.

Ru-biform. Having the form or nature of red.

Rubify. To redden.

Ruischiana Membrane. The third or chrio-capillaris membrane of the choroid.

Rutilant. Glittering, shining.

RULES

To find the deviating power of a prism, multiply the difference of index by the number of degrees of prism.

To find meter curve of a lens divide dioptric power by difference of index.

To find meter curve of a lens divide one meter by the radius of curvature on lens.

To find difference of index divide power of the lens by the meter curves of its radius.

Refractive power of a lens depends upon its curvature and the index of refraction of the glass combined. An increase of either one will produce greater refracting power. The shorter the focal length the greater the refractive power.

To find focal length of a lens surface divide the radius of curvature by difference of index.

To Convert—

Dioptries to meters of focal length, divide 1 by the number of dioptries.

Meters of focal length to dioptries, divide 1 by the number of meters.

Dioptries to centimeters of focal length, divide 100 by the number of dioptries.

Centimeters of focal length to dioptries, divide 100 by the number of centimeters.

Dioptries to inches of focal length, divide 40 by the number of dioptries.

Inches of focal length to dioptries, divide 40 by the number of inches.

To find dioptric value of any surface, multiply the difference of the index of refraction by the number of meter curves in the radius of curvature and give it the sign of the curve of the denser media. Remember two meter curves in optics mean one-half and three meter curves, one-third of a meter, and so on.

To find the angle of refraction, divide the angle of incidence by the index of refraction of the second media.

To find the radius of curvature of any media, multiply the focal length desired by the difference of the index in the two media.

Index of Refraction = Angle of Incidence divided by Angle of Refraction.

Angle of Incidence = Angle of Refraction multiplied by Index of Refraction.

Focal Length of Curved mirrors = one-half of the radius.

Power of a Mirror = one meter divided by its focal length (catoptries).

Metric Curve of a Mirror = one meter divided by the radius of the mirror.

Radius of Mirror = one meter divided by the metric curve of the mirror.

To find the dioptry power in any meridian of a cylinder, take the distance between the meridian of which you wish to know the power and the axis of the cylinder, multiply it by the power of the cylinder, and divide by 90.

To find the number of millimeters to decenter a lens for prismatic effect, multiply the prism wanted by 10 and divide by the power of the lens.

When prismatic effect is wanted in both the horizontal and vertical meridians, one prism can be used by placing a prism obliquely.

To find the prism to prescribe, square the power of the original prisms and add. Extract the square root of the sum, which will give you the power of the new prism.

To find the meridian to place the base of the new prism, divide 90 by the combined power of the original prisms and multiply by the vertical prism; this gives the distance from horizontal to place the base of the new prism.

To find the size of the image, focal length of emergent wave multiplied by size of object divided by focal length of incident wave; or. dioptric value of incident wave multiplied by size of object divided by dioptric value of the emergent wave.

To find the size of the object, reverse the formula above.

To find the size of the image on the retina, multiply the size of the object by the distance between the nodal points and the retina, then divide by the distance between the nodal points and the object.

To find circumference of a circle, multiply diameter by 3.1416.

To find diameter of a circle, multiply circumference by .31831.

To find area of a circle, multiply square of diameter by .7854.

To find area of a triangle, multiply base by one-half perpendicular height.

To find surface of a ball, multiply square of diameter by 3.1416.

RULES TO BE REMEMBERED

No. 1. No eye should be allowed to use accommodation at 20 feet or more.

No. 2. Always give a hyperope the strongest plus that will not blur his best distant vision.

No. 3. Give a myope the weakest minus that will give him best vision. Never put minus where it does not show returns.

No. 4. After putting the patient in the fog. place the axis of your minus cylinder at right angles to the plainest line seen.

No. 5. Correct presbyopia after correcting distant vision.

No. 6. Before testing for muscle trouble correct the ametropia.

SAC (L. saccus = a bag). A bag-like organ.

Saemisch's Ulcer (sa'-mish-ez). (Edwin Theodor Saemisch. Australian ophthalmologist, 1833.) Infectious corneal ulcer.

Sarcoma (sar-ko'-mah). (Gr. sarx = flesh + oma = tumor.) A tumor made up of a substance like the embryonic connective tissue. It is often highly malignant. Sarcoma of the ciliary body is generally pigmented, and often passes unobserved until it attains considerable size as a brown mass, which was at first concealed from view by the iris. Occasionally it makes its first appearance at the angle of the anterior chamber

Schematic Eye (ske-mat'-ik). (Gr. schema = shape, outline, plan.) A model or drawing that

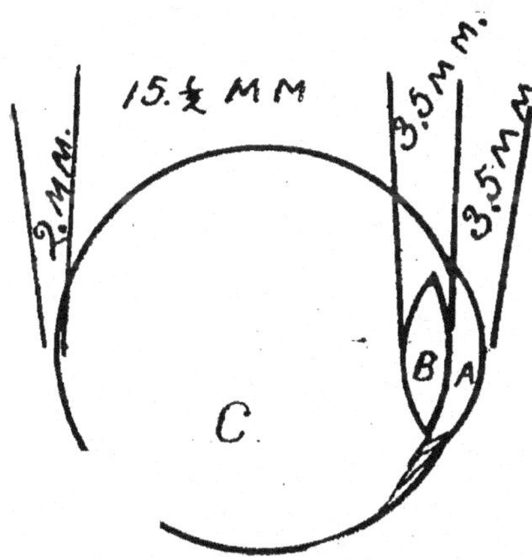

represents a normal or emmetropic eye. Used in demonstrating optical laws.

Schlemm's Canal (or circular venous sinus). (Friedrich Schlemm, German anatomist, 1795-1858.) A ring-like canal, of 0.3 by 0.5 mm. diameter, in the first tunic of the eye, between the cornea and the sclerotic. By means of the Spaces of Fontana it connects with the anterior chamber on one side, and directly communicates with the anterior ciliary veins on the other.

The Spaces of Fontana are formed by the dividing of the tissue from Descemet's Membrane in crossing from the corneal margin to pass into the base of the iris, and constitute the ligament pectinatum iridis.

Scintillation (scin-til-la'-shun). (L. scintilla = a spark.) A sensation of sparks before the eye.

Scissors Movement. A peculiar movement of the retinal reflex, resembling the opening and shutting of a pair of scissors. It indicates a condition of irregular astigmatism.

Sclera (skle'-ra). (Gr. skleros = hard.) The external and white coat of the eyeball, the sclerotic.

Scleral. Pertaining to the sclera.

Sclerectasia (skle-rec-ta'-si-ah). (Gr. sclera + ektasia = an extension.) A bulging state of the sclera.

Sclerectomy (skle-rek'-to-my). (Gr. skleros = hard + ektome = excision.) Excision of a portion of the sclera.

Sclerectasis (skle-rek'-ta-sis). A protrusion of the sclerotic coat. See Staphyloma.

Scleriritomy (skle-rir-it'-o-my). (Gr. skleros = hard + iris + tome = excision.) Incision of the sclera and iris in anterior staphyloma.

Scleritis (skle-ri'-tis). (Gr. skleros = hard + itis = inflammation.) Inflammation of the sclerotic coat.

Sclerochoroiditis (skle-ro-cho-roid-i'-tis). Inflammation of both the choroid and the sclerotic coats of the eye.

Scleroconjunctival (skle-ro-con-junc-ti'-val). That condition in which the sclera and the conjunctiva are concerned.

Sclerocorneal Sulsus (furrow). The angle or depression formed by the difference in the radius of curvature of the sclerotic and cornea. This angle makes the eyeball stronger and more firm at this point, and it is just inside this angle that the ciliary muscles are attached.

Sclerocorneal (skle-ro-cor′-ne-al). Relating to the sclerotic coat and cornea.

Scleroiritis (skle-ro-i-ri′-tis). An inflammation which involves both the iris and the sclera.

Sclerokeratoiri′tis. Inflammation of the sclera, cornea, and iris.

Scleronyxis (sklĕ-ro-nyx′-is). (Gr. skleros = hard + nyxis = a pricking.) A perforation of the sclerotic coat.

Sclerophthalmia (skle-rof-thal′-mi-ah). (Gr. skleros = hard + ophthalmos = eye.) That condition in which the sclera overlaps the cornea, so that only a portion of the latter remains clear.

Scle′rosed (Gr. skleros = hard). That condition in which a part is affected with sclerosis; a hardening.

Sclerosis (sclero′-sis). (Gr. sklerosis = hardness.) The process of becoming hard, tough, or indurated.

Sclerotic (skle-rot′-ic). (Gr. skleros = hard.) The posterior five-sixths of the first tunic. It is firm, hard, and opaque; known as the white of the eye. It serves to give shape to the globe, protects its more delicate interior, and at the same time acts as a dark-box or camera. It is to this coat that the muscles are attached. The sclerotic is thickest in the posterior part, where it has a thickness of about 1 mm. It gradually diminishes in thickness toward the anterior part, becoming somewhat thicker near the cornea, because here the tendons of the recti muscles are attached and fused with it. The sclerotic consists of fine cotton-like fibers or connective tissues, which are united into bundles which seem to be woven in all directions. Between the bundles are found lymph-spaces, which are in part lined with fat cells. The sclera has very few blood-vessels and nerves. The blood-vessels are derived from the anterior ciliary and posterior ciliary arteries. The venous blood is removed by the venae vorticosae and the anterior ciliary veins. Its nerves are derived from the ciliary nerves.

Scleroticectomy (skler-ot-i-kek′-to-my). (Gr. skleros = hard + ektome = excision.) An operation for artificial pupil by removal of a portion of the sclerotic.

Sclerotomy (skle-rot′-o-me). (Gr. skleros = hard + tome = incision.) Surgical incision of the sclera.

Scotodinia (sko-to-din′-iah). (Gr. skotos = darkness + dine = a whirling.) Dizziness, with headache and dimness of vision.

Scotoma (sko-to′-mah). (Gr. skotoma = darkness.) That condition in which there is a blind or partially blind area in the visual field. Sometimes the patient will complain of seeing dark, vanishing, cloudy spots before the eyes. **Absolute S.**, a part of the visual field in which there is absolute blindness.

Scotometer (sko-tom′-e-ter). An instrument for measuring scotoma.

Sebaceous Cysts (L. sebum = suet + Gr. kystis = bladder.) A small rounded body, the size of a pea, which appears in the thicker portions of the skin of the eyelids.

Seborrhea (seb-or-e′-ah). (L. sebum = suet + Gr. rhola = a flow.) An abnormal secretion of the sebaceous glands.

Se′cant (L. secare = to cut). (Geometry.) A line that cuts another, or divides it into parts. The secant of a circle is a line drawn from the circumference on one side to a point on the outside of the circumference on the other.

Secondary Axis. See Axis.

Secondary Foci. See Focus.

Seg′ment (L. segmentum; secare = to cut). A section of a circle. A cylindrical lens is a segment of a cylinder which refracts rays of light in all meridians but one. This meridian is known as the axis. A spherical lens is a segment of a sphere. A segment of anything is one of the parts into which it is divided.

Serpiginous (ser-pij′-in-us). (L. serpere = to creep.) Resembling a ringworm.

Shadow Test. See Retinoscopy.

Sheath. A tubular case or envelope. **Optic S.**, the covering of the optic nerve formed by the dura mater on the outside, and the pia mater on the inside, of the subarachnoid space.

Shortsightedness. See Myopia.

Sight. The sense by which external objects are located and seen, their color, size, and form described, as compared with other objects, through the medium of the visual organ.

Sign. That by which anything is represented. The sign of addition (+) represents convex sperical and convex cylindrical lenses. The sign of subtraction (—) is used to represent concave spherical and concave cylindrical lenses.

Sine. (L. sinus = sine.) The length of a perpendicular drawn from one extremity of an arc of a circle to the diameter drawn through the other extremity. Sine of an angle is a circle whose radius is unity, the sine of the arc that measures the angle; in a right-angled triangle, the side opposite the given angle divided by the hypotenuse. Versed sine, that part of the diameter between the sine and the arc.

Sinis'trad. (Gr. sinister = left + ad = to.) To or toward the left.

Sinus (si'-nus). (L. "gulf.") A hollow cavity. **Frontal Sinus,** one of the two irregular cavities in the frontal bone containing air and communicating with the nose through a funnel-shaped passage. **Occipital Sinus** is the smallest of the cranial sinuses, occasionally there are two. It is situated in the attached margin of the falx cerebelli, opening into the torcular Herophili. **Lateral Sinus,** two veins of the dura running along the crucial ridge of the occipital bone. **Cavernous Sinus,** venous cavities, starting behind the sphenoidal fissure, running back on the side of the pituitary fossa, and joining the superior and inferior petrosal sinuses. Each cavernous sinus receives anteriorly the superior ophthalmic vein through the sphenoidal fissure, on the inner wall of each sinus is the internal corotid artery, filaments of the corotid plexus, abducent nerve; and on the outer wall, the trochlear, ophthalmic, oculo motor and the maxillary division of the trigeminal nerves.

Skiascope (ski′-as-kope). (Gr. skia = shadow + skopeo = I examine.) Better known as the retinoscope.

Skiascopy. (Gr. skia = shadow + skopeo = I examine.) See Retinoscopy.

Snellen, M. D., Prof. H. Born in Holland. A professor of ophthalmology occupying the chair at the University at Utrecht, Holland. A pupil of Dr. F. C. Donders, whom he succeeded in practice and his professorship. He devised a chart consisting of letters and symbols by which the subjective means of measuring the range of vision could be uniformly and scientifically determined. It is held by Snellen that in order to distinguish one letter from another the eye must be able to distinguish the spaces between the lines which correspond to a visual angle of 1′. This is true for certain letters, as, for instance, to differentiate between O and C, where the eye

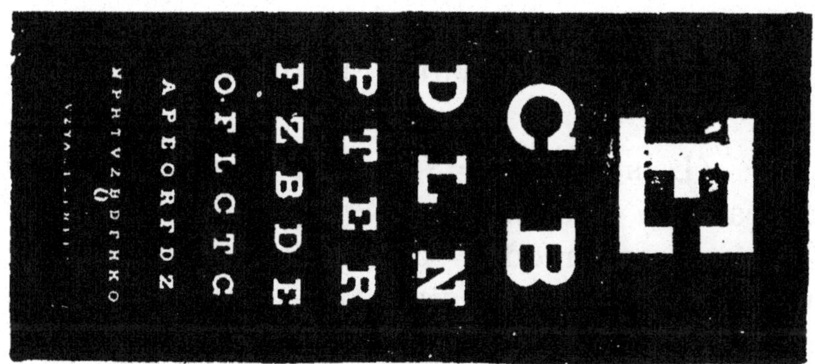

Snellen Chart.

must distinguish the white space which interrupts the circle in C. The same is true for E and F, but the principle is not applicable to the other letters of his series. In a lecture on refraction by Landolt, we learn of what great advantage it is to determine the visual acuteness and the refraction at the same time. We must determine the refraction at such a distance as shall exclude the accommodation as much as possible. For this a distance of twenty feet, or six metres, is necessary. We therefore place our test type at 20 F, and see what are the smallest characters which each eye, separately, can distinguish. These types are so designed that at the distance at which they should be seen they each subtend an angle of 5′ at the eye. And when the letters marked 20, or 6 M, are read from 20 F, vision is said to be normal, and an eye with normal vision can read any of the letters on the chart at the distance at which they are marked.

Snow-blindness. Long exposure of the eyes to the glare from snow gives rise to an acute conjunctivitis, attended with intense pain, photophobia, and occasionally conjunctival hemorrhages.

Socket (sok′-ket). (L. soccus = a show. a sock.) A hollow part into which a corresponding part fits.

Spasm. (L. spasmus = I draw.) An involuntary contraction of a muscle. Spasm of accommodation is a spasmodic contraction of the ciliary muscles, thus increasing the convexity of the crystalline lens and making the eye appear to have a higher refractive power. There are two kinds—tonic and clonic. **Tonic s.** is where the spasm persists for a considerable time, and **Clonic s.** is where the muscles contract and relax intermittently.

Spectacles. (L. spectare = to regard.) A pair of lenses mounted in frames with temples attached.

Spectrum (spec′-trum). White light is composed of all the colored lights known, and when it is separated by a prism or other means and thrown on a white screen, in an otherwise dark room, a band of colors resembling a rainbow is seen. This is called a prismatic or solar spectrum.

Of the seven primary colors which form the spectrum, Violet is refracted the most, then Indigo, Blue, Green, Yellow, Orange and Red the least. There are also invisible rays called "ultra"-red, and "ultra"-violet beyond its apparent boundaries.

Ocular Spectrum color seen by an eye where none exists.

Diffraction Spectrum is a spectrum produced by diffraction.

Chromatic Spectrum is the visible colored rays of the solar spectrum, showing the seven principal colors in their order and covering the larger portion of the space of the whole spectrum.

Sphenoid (sphe′-noid). (Gr. sphen = wedge + eidos = resemblance.) Sphenoid Bone. A bone situated at the upper and back part of the orbits on the median line, at the base of the cranium. It articulates with all the other bones of that cavity, and strengthens their union. When seen from above it resembles a bat with its wings extended.

Sphere (sfer). (Gr. sphaira = a ball.) A ball-like body.

Spherical. Having the form of a sphere. **Spherical Lens** is one, the curved surface of which is a segment of a sphere and is known as a lens with the same refracting power in all its meridians. There are three ways to grind a plus or minus sphere of the same value; namely, bi-concave, plano-concave, periscopic-concave, bi-convex, plano-convex, periscopic-convex. See Lenses.

Spheroid (sphe′-roid). (Gr. sphaira = sphere + eidos = resemblance.) That which resembles a sphere in shape.

Spherometer. (Gr. sphaira = sphere + metron = measure.) An apparatus for measuring the curvature of a surface.

Sphincter (sphinc′-ter). (Gr. sphinkter = a band.) A ring-like muscle. The sphincter muscle of the iris when contracted closes down the pupil. When relaxed allows the pupil to become dilated.

Spintherism (spin′-ther-ism) (Gr. spinther = spark.) That condition in which the patient complains of seeing star-like flashes of light.

Squamous (skwa′-mus). Scaly.

Square (skwar). (L. quatuor = four.) (a) An equilateral rectangle. (b) The second power of a number. (c) To raise a number to the second power.

Squint. (Fr. guigner to wink or direct with one eye.) The act of half closing the eyelids while viewing an object. The word squint is sometimes used to denote strabismus.

Staphyloma (sta-fy-lo′-mah). (Gr. staphyle = grape + oma = tumor.) A bulging of the cornea or sclera. **Anterior s.**, a bulging forward of the anterior portion of the eye. **Posterior s.**, backward bulging of the posterior pole of the eye.

Stat'ic.. (Gr. statikos = causing to stand.) Not in motion; in a state of rest. The static retraction is the refraction of the eye with the muscles of accommodation at rest; just the reverse to dynamic refraction.

Steato'sis. (Gr. stear (steat) = tallow + suffix osis = condition.) That condition in which we have fatty degeneration; disease of the sebaceous glands.

Stenopaic Slit (sten-o-pa'-ic). (Gr. stenos = narrow + ope = opening.) An accessory to be found in any complete trial case, and consists of an opaque disc with a slit about an inch long

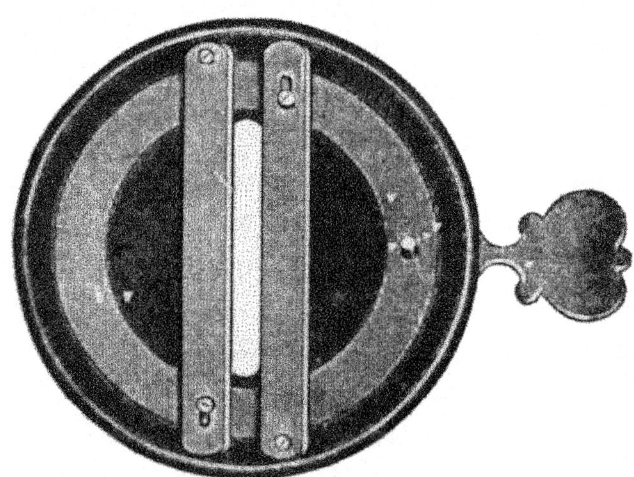

and one millimetre wide. It is used for the purpose of finding the principal meridians in cases of regular astigmatism. If vision is near normal it is best to fog or blur it a line or two with a plus sphere, then place the disc in the trial frame in front of the eye we are about to examine, while the other is covered by the opaque disc. Instruct the patient to revolve the disc in the trial frame while trying to see the letters on Snellen's test type at a distance of twenty feet (never the astigmatic wheel) and stop when the best vision is obtained, thus locating the principal meridian with least error. We then note the degree mark on the trial frame to which the slit is pointing and make a right angle cross on a piece of paper showing this meridian and the one at right angles. We now proceed to correct the error by placing spheres over the slit until we find the strongest plus or weakest minus that allows the best vision, writing the amount on the arm of the cross corresponding to the slit. Then revolve the slit 90° and fit this meridian as above, writing the amount on the other arm of the cross. From this cross we write the Rx without any change. This will represent the Ametropia.

Stereoscope (ster′-e-o-scope). (Gr. stereos = solid + skopeo = I view.) An instrument composed of two prisms arranged in such a way that two separate pictures of the same kind may be seen as one. This instrument makes the picture more natural, as the objects appear to stand out.

Stereoscopic Vision (ster-e-o-scop′-ic). Where we have equal vision with the two eyes and the objects appear to stand out in solid form, and are not seen as flat pictures.

Stilling's Canal. (Benedict Stilling, German anatomist, 1810-1879.) A small canal leading from the optic disc through the vitreous humor to the lens of the eye. See Anatomy.

Stillicidium (stil-li-sid′-i-um). (L. stilla = drop + cadere = to fall.) An overflowing of the tears upon the check due to a stricture of or a narrowing of the nasal duct. Same as epiphora.

Stilus (sti′-lus). (L. stilus = a stake.) A small instrument made of gold or silver used for dilating the lacrimal duct.

Stoke's Lenses. (William Stokes. Dublin physician, 1804-1878.) An instrument that was used in the diagnosis of astigmatism.

Stop-needle. A needle with a disc attached to regulate the depth of penetration.

Strabismometer (strab-is-mom′-e-ter). (Gr. strabismos + metron = measure.) An instrument for measuring the degrees of strabismus.

Strabismus (stra-bis′-mus). (Gr. strabismos = distorted.) (Cross-eyed.) That condition in which the eyes are not parallel for distant vision. The visual axis of one eye only is directed towards the object looked at; this is known as the fixing eye, while the other is known as the deviating eye. It is caused by anything which develops preponderance of power in a muscle, either directly or indirectly. It may be due to an uncorrected error of refraction; or from anything which prevents binocular vision, such as cataract, corneal opacities, displaced macula lutea, a short, long, or paralyzed muscle. However, it is well to **correct any ametropia,** for when a person has an error of refraction in one eye that interferes with the vision of its fellow, he will learn to turn the eye with the error to one side. **Alternating s.,** affecting both eyes equally, but not at the same time. **Concomitant s.** is that form of strabismus in which one eye, although deviated, always moves with the other, so that the

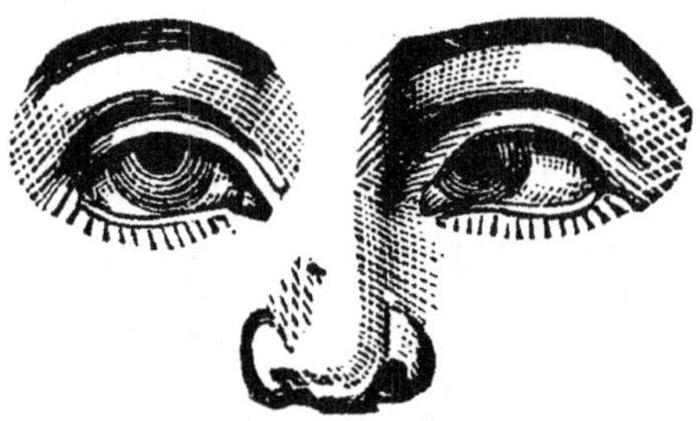

One Eye turning in.

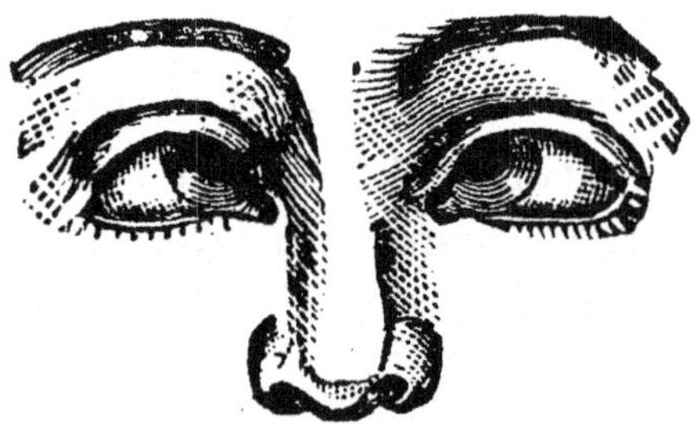

Both Eyes turning in.

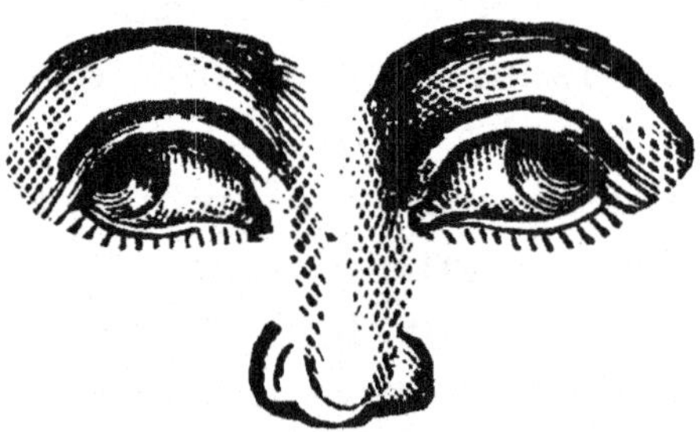

Both Eyes turning out.

amount of deflection remains the same. **Paralytic s.** is due to paralysis of one or more of the extrinsic muscles, and the eye remains stationary. Hypermetropia is responsible for 80 per cent of converging strabismus on account of the ciliary muscles and the internal rectus muscles being supplied by one and the same nerve.

When the eye attempts to accommodate in order to overcome the hypermetropia, the internal rectus will contract, and if the patient has not the nerve energy to control the external rectus, the eye will turn in. See Heterotropia.

Strabotomy (stra-bot′-o-my). (Gr. "oblique" + tome = a cutting.) An operation calling for the cutting of an ocular tendon for relief in cases of strabismus.

Strain. (L. stringere = to bind.) Injury from over-use. **Ciliary s.**, the result of overwork of the ciliary muscles in hypermetropia and sometimes in emmetropia. **Muscular s.**, overwork of the extrinsic muscles as in heterophoria. **Retinal s.**, fatigue of the retina caused by too strong light or from over-use in a normal light. The eye should be protected from all direct rays of light, as only reflected light is necessary for vision.

Stroma (stro′-ma). (Gr. "I spread out.") The foundation tissue or support of a formation.

Stye ("to rise") or **Hordiolum.** A small boil affecting the connective tissue near the edge of the eyelid, sometimes several appear at once, or there may be a succession of them. They cause swelling of the lid. In a day or two the swelling increases, with considerable pain, and the skin over it becomes red, afterwards showing a yellowish discoloration at the center that finally opens near the border of the lid, with a discharge of pus. After which the inflammatory symptoms abate and the cavity soon closes.

Treatment: Small poultices or hot fomentations until pus forms, then open by incision parallel to the edge of the lid. General health requires attention if eye strain is not the cause.

Subarach'noid Space. (L. sub = under + Gr. arachne = cobweb + e i d o s = resemblance.) That space between the dura mater and the pia mater which forms the optic sheath and the Capsule of Tenon.

Subconjunctival (sub-con-junc-ti'-val). That which is situated just beneath the conjunctiva.

Subjective (sub-jec'-tive). That which pertains to, or is perceived by, an individual. Not perceptible to any other person. It refers to the patient as he sees objects or feels concerning his own impressions.

Subla'tio Ret'inae. (L. sublatus = taken away + rete = net.) Detachment of the retina.

Subluxa'tion. (L. sub = under + luxare = to dislocate.) Where the lens of the eye is a little displaced, subluxation may consist in the lens being turned a little obliquely, so that one end of it looks somewhat forward. This condition may be recognized from the unequal depth of the anterior chamber. In cases of luxation, that is, where the lens has left its place in its capsule, so that it partly covers the pupil, that part of the pupil which is a deep black has no lens, while the part which contains the lens would be of a delicate gray. Any dislocation of the lens entails a considerable disturbance of vision. If the lens still lies behind the pupil the eye becomes very myopic, because the lens is allowed to assume its maximum convexity on account of separating from the suspensory ligaments which keep it elongated when the eye is at rest. Invariably any tear in the suspensory ligament results in soft cataract. Added to this is a considerable degree of astigmatism. Dislocations of the lens usually entail secondary consequences which may be extremely disastrous to the eye, but in those cases in which the dislocation entails no further injurious consequences than the disturbance of vision, the treatment consists in prescribing suitable glasses.

Suborbital (sub-or′-bit-al). Beneath the orbit.

Subretinal (sub-ret′-in-al). Situated beneath the retina.

Subtraction (sub-trak′-shun). (L. sub = under + trahere = draw.) The operation of finding the difference between two numbers.

Subvolution (sub-vo-lu′-shun). (L. sub = under + volvere = to turn.) An operation for the removal of a pterygium.

Suction (suc′-shun). (L. sugere = to suck.) A method by which fluid is withdrawn.

Suffusion (suf-fu′-zhun). L. suffundere = to pour down.) State of being blood-shot, or of being moistened. A suffusion of tears is an excess of the flow of tears.

Super Cilia. (Upper hairs.) (L. super = above + cilium = eyelash.) The eyebrows.

Superciliary (su-per-cil′-i-a-ry). That which pertains to the eyebrow.

Supra Choroidal Space. The space between the sclerotic and choroid.

Supraduction, Sursumvergence. (L. sursum = upward + vergere = to bend. The act or power of turning one eye above its fellow.

Supraorbital (su-pra-or′-bi-tal). (L. supra = above + orbita = orbit.) Located over the orbits.

Supra-orbital Foramen. A small passage in the Supra-orbital Ridge through which passes the supra-orbital nerve (a branch of the fifth) artery and vein.

Surface (ser′-fas). (L. superficies = the upper face.) The bounding or limiting parts of a solid.

Sursumduction (sur-sum-duk′-shun). (L. sursum = upward + ducere = to draw.) The act of turning one eye upward independent of its fellow. The test is made by placing the base of the prism down until we find the strongest which the eyes can see an object singly. It is seldom more than 3°

Sursumvergence (sur-sum-vur′-jenz). (L. sursum = upward + vergere = to turn.) An upward turning of the eye.

Sursumversion (sur-sum-vur′-shun). (L. sursum = upward + vertere = to turn.) The act of turning the eyes upward.

Suspensory (sus-pen′-so-ry). (L. suspendere = to suspend.) Serving to hold up a part.

Suspensory Ligaments. The hyaloid membrane forms the hyaloid sac in which the vitreous humor is contained. It runs forward up over the ciliary body, divides and forms the suspensory ligaments, which are attached to the lens capsule. C. B. Lockwood, in a journal of Anatomy and Physiology, vol. XX., part I.—Ed. of

15th English edition, has also described a thickening of the lower part of the Capsule of Tenon, which he has named the suspensory ligament of the eye. It is slung like a hammock below the eyeball, being expanded in the center and narrow at its extremities, which are attached to the malar and lachrymal bones respectively.

Suture (su′-ture). (L. sutura = a seam.) The serrated junction of the intracranial bones. Dovetail joint.

Sylvius, Aqueduct of. A passage from the third to the fourth ventricle of the brain.

Symblepharon (sym-blef′-ar-on). (Gr. syn = together + blephron = eyelid.) Adhesion of the lids to the eyeball. This develops whenever two opposed spots of the conjunctiva of the lid and of the eyeball have raw surfaces which come into contact with each other, and in consequence become adherent. Causes which can give rise to the formation of raw surfaces upon the conjunctiva are burns by the action of heat, burns from caustic substances, diphtheria, operations, ulcers of all kinds, etc.

Sympathetic Ophthalmitis (sym-pa-thet′-ik of-thal-mi′-tis). (Gr. syn = with + pathos = suffering + ophthalmos = eye + itis.) An inflammatory condition of the iris and ciliary body, which is developed through an injury or disease of the opposite eye.

Symptoms. (Gr. syn = with + ptoma = I fall.) A perceptible change which indicates disease, or that which indicates the existence of something else. See Objective and Subjective Symptoms.

Synchysis (syn′-chy-sis). (Gr. confusion.) Liquefaction of the vitreous. When observing opacities of the vitreous with the ophthalmoscope, we see that most of them float about freely in the vitreous. From this we would assume that the framework of the vitreous must have been destroyed, so that this body itself is converted into a perfectly liquid mass.

Syndesmi′tis. (Gr. syndesmos = ligament + itis = inflammation.) That condition in which there is inflammation of a ligament or of the conjunctiva.

Synechia (syn-e′-chi-ah). (Gr. synecheia = continuous.) Adhesion, as of the iris to the lens or cornea. **Posterior s.,** adhesions of the iris to the lens capsule. **Anterior s.,** adhesions of the iris to the cornea.

Synizesis (sin-iz-e′-sis). (Gr. "a falling in.") Contraction of the pupil of the eye.

Synophthalmus (syn-of-thal′-mus). (Gr. syn = together + ophthalmos = eye.) A one-eyed monster.

Syntropic (sin-trop′-ik). (Gr. syn = together + tropikos = turning.) Turned in the same direction.

System. (Gr. systema = to place together.) A bodily organism. An assemblage of parts or organs which unite in a common function.

T. An abbreviation for tension or temperature.

Tangent (tan′-gent). (L. tangere = to touch.) Touching at a single point; specifically meeting a curve or surface at a point and having at that point the same direction as the curve or surface—said of a straight line, curve or surface; as, a line tangent to a curve; a curve tangent to a surface; tangent surfaces. Tangent plane is a plane which touches a surface in a point or line.

Tapetum (ta-pe′-tum). (L. "a carpet.") The luminosity seen in the eyes of many beasts. A lustrous, greenish membrane seen in the eyes of cats and many animals that require night vision.

Tarsal Cartilages (tar′-sal kar′-til-aj-es). (Gr. tarsos = a wicker work frame + cartilago = gristle.) That which forms the tough skeleton layer of the eyelids, giving them rigidity of form and affording them firm support. The shape is like that of the lids being fastened around the edge of the orbit. The tarsus of the upper lid is broader than that of the lower.

Tarsitis (tars-i′-tis). (Gr. tarsos = a wicker work frame + itis = inflammation.) An inflammation involving the tarsal cartilages.

Tarsoplasty (tar′-so-plas-ty). (Gr. tarsos + plasso = I form.) Plastic surgery of the tarsus.

Tarsorrhaphy (tar-sor′-a-fe). (Gr. tarsos + rhaphe = a stitching.) An operation upon the eyelids.

Tarsotomy (tar-sot′-o-my). (Gr. tarsos + tome = incision.) A surgical operation which involves the cutting of the tarsal cartilages.

Tarsus (tar′-sus). (Gr. tarsos = a wicker work frame.) That which forms the skeleton of the eyelid, giving it rigidity of form and affording it firm support. The tarsus of the upper lid is broader than that of the lower.

Tears. The watery secretion of the lacrimal glands.

Teichopsia (tei-kop'-si-ah). (Gr. teichos = wall + opsis = vision). A luminous appearance before the eyes, with a zigzag, wall-like outline.

Telangiectasis (tel-an-je-ek'-ta-sis). (Gr. tĕlos = end = angeion = vessel + ektasis = a stretching out.) Dilatation of capillaries.

Tendency (tend'-en-cy). A disposition on the part of a muscle to incline toward certain directions.

Tendon (ten'-don). (L. tendo = I stretch.) The fibrous cords by which the muscles are attached.

Tendons. (Sinew.) White, glistening, fibrous cords, varying in length and thickness, sometimes round, sometimes flattened, of considerable strength and devoid of elasticity. Aponeuroses are flattened or ribbon-shaped tendons. They are without nerves and have very few blood-vessels. Tendons pass through all muscles and form their attachment at each end.

Tendon of Lockwood gives origin to the Superior Internal and upper head of the External recti muscles. It is part of the ligament of Zinn.

Tenonitis (ten-on-i'-tis). (Gr. tenon = tendon + itis.) Inflammation of the Capsule of Tenon.

Tenon's Capsule. (Jacques Tenon, French anatomist, 1724-1816.) (Tunica vaginalis oculi.) The cup-like thin membranes which envelop the eyeball, covering the sclera from the optic nerve to the ciliary region, where it joins the ocular subconjunctival tissue. The space within the orbit which is not occupied by the eyeball, its muscles, nerves, vessels or other parts belonging to it is completely filled with soft, fat and delicate, elastic connective tissue. In various places this tissue is condensed into two layers of considerable strength. One layer investing the sclera of the eyeball and the other lining the cushion of fat in which the eyeball rests. These two layers lined with flattened endothelial cells encloses a lymph space which communicates with the subdural and subarachnoid lymph spaces of the optic sheath. They are traversed by delicate bands of connective, elastic tissue, thus forming a flexible socket, in which the eyeball rotates by means of its muscles. This capsule is perforated by the ocular muscles and is reflected back on each as a tubular sheath. These two layers are sometimes referred to as the **dura mater** and **pia mater,** owing to their connection through the optic sheath, with the dura and pia mater which lines the skull.

Tenotomy (ten-ot′-om-e). (Gr. tenon = tendon + tome = incision.) An operation for cutting or dividing the tendon of a muscle.

Tension (ten′-shun). (L. tendere = to stretch.) The condition of being stretched or tense.

Tensor-tarsi Muscle (ten-sor-tar′-si). A very small muscle located at the inner canthus of the eye. It takes its origin at the crest of the lacrimal bone, and is inserted into the tarsal cartilage of the eye-lids. It is supplied by the facial nerve. Its use is to compress the puncta and lacrimal sac.

Test. (L. testis = witness.) An examination or

trial. T. **types,** letters of various shapes and sizes used in testing visual power.

Tetranopsia (tet-ran-op′-si-a). (Gr. tetra = four + an = away + opsis = vision.) **Obliteration** of one-fourth of the visual field.

Thermometer (ther-mom′-e-ter). (Gr. therme = heat + metron = measure.) An instrument by which temperature is measured.

Thalamus (thal′-a-mus). (A room; a bed.) The place in which a nerve originates, **Optic Thalamus.** A mass of nerve matter on both sides of the third ventricle of the brain.

Thrombosis (throm-bo′-sis). (Gr. thrombosis = a curdling, a clot.) The formation of a blood-clot in a vessel at the point of obstruction.

Thyroid (thi′-roid). (Gr. thyreos = an oblong shield + eidos = form.) Shield-shaped. T. **Gland** is a vascular body situated at the front and sides of the neck, and extending upwards upon each side of the larynx. It is a single gland, varying greatly in size in different individuals.

Tinea Tarsi. Blepharitis marginalis. See Blepharitis.

Tobacco Amaurosis. (Gr. amauros = obscure.) A dimness of vision caused by the excessive use of tobacco, which acts directly upon the nervous system. The reduction in the visual acuity is almost always the same in both eyes. Treatment consists, first of all, in abstinence from tobacco, and it is probable that in light cases this alone is sufficient to effect a cure.

Tonic Spasm. (Gr. tone + "to draw.") A continuous involuntary contraction of the ciliary muscles. This condition may exist in any muscle.

Tonometer (to-nom′-e-ter). (Gr. tonos = tone + metron = measure.) An instrument for measuring the tension of the eyeball.

Toric Lens. A lens with a sphere and a cylinder on the same side, usually periscopic in shape. See Lens.

Torsion (tor′-shun). (L. torquere = to twist.) A twisting.

Toxic Amblyopia. (Gr. toxikon = poison.) Amblyopia caused by a poison, a common cause being excessive use of tobacco or liquor or both.

Trachoma (tra-ko′-mah). (Gr. trachys = rough.) Granular conjunctivitis. Characterized by slowly progressive changes in the conjunctiva of the eyelids, in consequence of which this membrane becomes thickened, vascular, and roughened by firm, round elevations, instead of being pale, thin and smooth. Granular disease is very important. because it greatly increases the susceptibility of the conjunctiva to take on acute inflammation and to produce contagious discharge. It often gives rise to deformities of the lid and to serious damage of the cornea. The conditions which favor the development and spread of trachoma are unclean and overcrowded surroundings in which ventilation is neglected. and the locality is damp. The disease is common among school children who are poorly nourished.

Tract. See Optic Tract.

Transection (tran-sek′-shun). (L. trans = across

+ sectio; secare = tc cut.) A section mad across a long axis.

Transillumination (trans-il-lu-min-a′-shun). (L trans = through + illuminare = to light up. The inspection of the interior of an organ by means of a strong light.

Transit. (L. trans = through.) A passing across A term used in retinoscopy to indicate movement of the light area.

Transi′tional Zone. The posterior part of the lens sac during the stage of growth.

Translu′cent. (L. trans = through + lucere = to shine.) The quality of transmitting rays of light without the object being distinctly seen. (Frosted glass.)

Transparent. (L. trans = through + parere = to appear.) Having the property of being clearly seen through.

Transposition (trans-po-si′-shun). (L. trans = across + ponere = to place.) Changing the form of an optical prescription without changing its optical value To transpose a lens is to change its curves without changing its refractive value.

To transpose an optical prescription is to change the form or shape of the lens without changing its optical value, periscopic effect being the prime and important feature in most instances. The term periscopic is applied to lenses having concavo-convex surfaces, which enable the eye to view with equal likeness on all sides. When lenses are not of this description the desired result may be obtained by the following rules:

When the sign of the sphere and cylinder are alike (i. e., both plus or both minus) add them together for the new sphere, prefixing the same sign.

When the sign of the sphere and cylinder are different (i. e., one plus and the other minus) subtract for the new sphere, prefixing the sign of the larger number.

Always change the sign of the cylinder to the opposite, but do not change its value.

Always change the axis to right angle (i. e., move it 90°).

For transposition of simple cylinders, use the following rule: Use the numerical value of your cylinder for the new sphere, prefixing the same sign, and for the new cylinder use the same value as the original, but prefix the opposite sign and change the axis to right angle.

To convert cross cylinders into sphero-cylinders, apply the following rule: Use the smaller number for your sphere (if the numbers are alike, take either one, keeping its own sign), and when the signs of the cylinders are alike (i. e., both plus or both minus) subtract them for your cylinder, prefixing the same sign. When the signs are unlike (i. e., one plus and the other minus) add them for your cylinder, prefixing the sign of the remaining cylinder, and also its axis.

If, after transposing cross cylinders, your prescription is not periscopic, make it so by transposing again by one of above rules.

Below find a few examples in transposition, with their answers:

Example, + 3 sph. ⌢ — 2 cyl. ax. 60.
Answer, + 1 sph. ⌢ + 2 cyl. ax. 150.
Example, + 2 sph. ⌢ + 2 cyl. ax. 90.
Answer, + 4 sph. ◯ — 2 cyl. ax. 180.
Example, + 4 cyl. ax. 45 ◯ + 2 cyl. ax. 135.
Answer, + 2 sph. ⌢ + 2 cyl. ax. 45.
Example, — 3 cyl. ax. 20 ◯ + 3 cyl. ax. 110.
Answer, — 3 sph. ⌢ + 6 cyl. ax. 110.
Example, + 1 cyl. ax. 60.
Answer, + 1 sph. ◯ — 1 cyl. ax. 150.

Writing a prescription from a cross is not transposing. We must first have a written prescription before it can be transposed.

In order to give the patient glasses which give them the best possible results, it will be necessary to know how to build lenses of different shapes, for instance:

Biconvex, biconcave, plano convex, plano concave, periscopic and toric.

Lenses have two kinds of power, minus and plus—the former being thinner in the center and the latter thinner at the edge. These lenses can be made up as a sphere or cylinder.

A sphere is a lens with the same power in all its meridians.

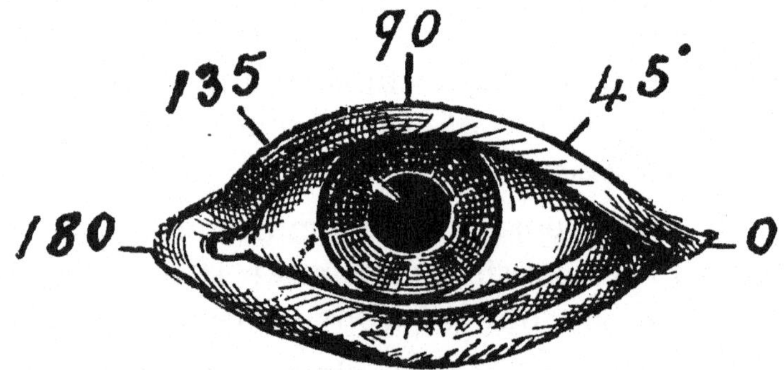

Cut showing how the meridians of an eye are numbered from right to left.

A meridian is any straight line drawn from edge to edge over its optical center.

The optical center being a point in line with the thickest part of a plus and the thinnest part of a minus lens.

A cylinder is a lens with power in all meridians but one, this one, having no power and is called its axis. The full power of a cylinder is always found at right angles to its axis.

In the following diagram we will use a plus four dioptry sphere and plus four cylinder for example:

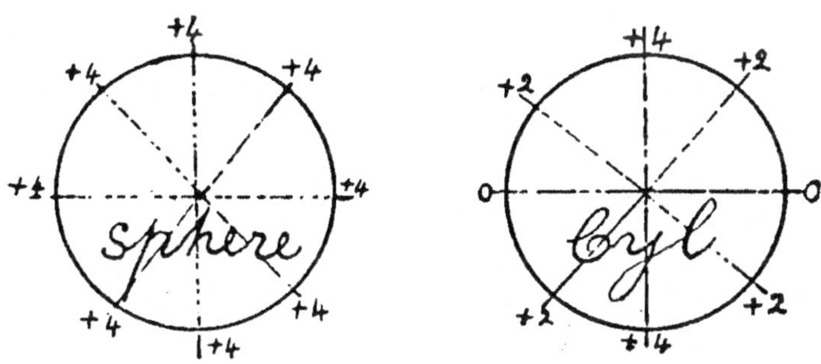

Notice that the power is the same in all meridians of a sphere, while those of a cylinder vary in power.

An optical prescription is nothing more than an order for a lens of a given power and shape. and when it is transposed, the shape is changed but not its optical value (or power); for instance, we take the following prescription:

+ 4 sph. ⌣ + 4 cyl. ax. 180.

which reads plus four sphere combined with a plus four cylinder, axis 180. The optician, on receiving this prescription, will grind the plus four sphere on one side of the lens and a plus four cylinder on the other and cut it out. so that the axis of the cylinder will be at 180° This lens being plus on both sides is known as a biconvex lens.

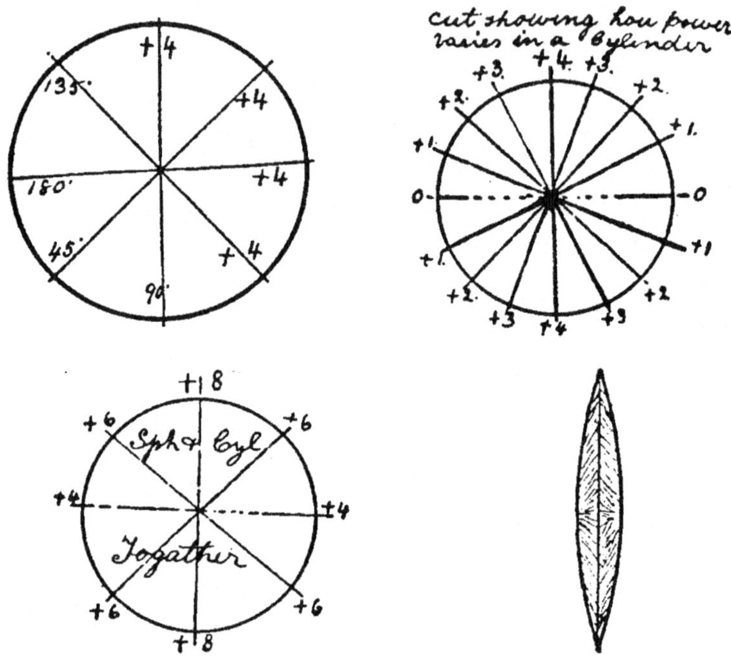

In this example we have the sphere and **cylinder** separated and together, showing their combined powers and also their appearance from the side. It should be noted that the sphere does not change its value under the axis of the cylinder, thus forming one of its principal meridians.

In order to change the shape of this lens we must apply the following rule:

When the signs of the sphere and cylinder are alike, that is, both plus or both minus, add the values together for your new sphere which would be plus eight, then change the sign of your cylinder, which makes it minus, but do not change its value. Change its axis 90°, taking 90 from 180 leaves 90, thus + 4 sph. ⌢ + 4 cyl. ax. 180 transposed gives you + 8. sph. ⌢ — 4. cyl. ax. 90°.

In the latter prescription you have what is known as a periscopic lens, one side plus and the other side minus.

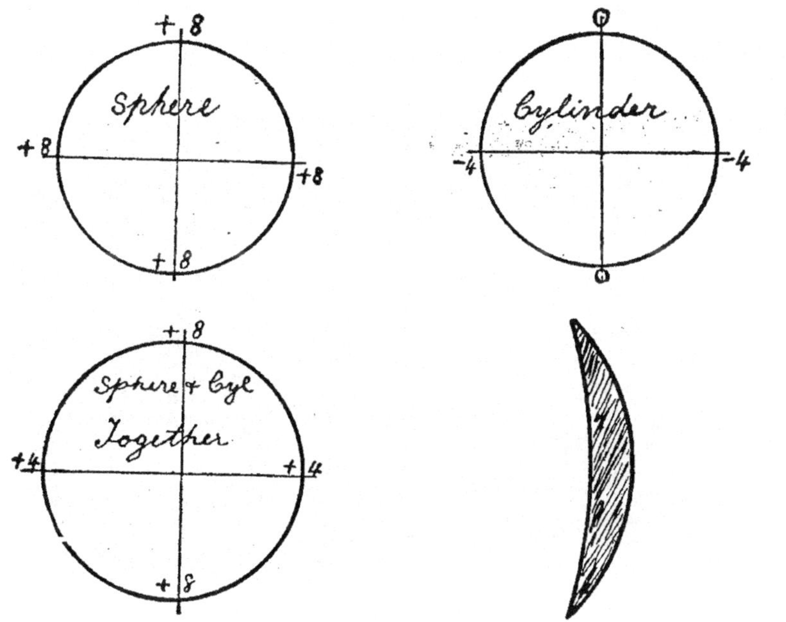

This shape lens is much preferred by the Refractionist of today on account of its appearance and comfort to the patient.

Trapezoid (tra-pe′-zoid). (Gr. trapeza = table + eidos = form.) A quadrilateral having two parallel sides.

Traumatic (trau-mat′-ik). Of, caused by, or pertaining to, an injury.

Trembling Eyes. See Nystagmus.

Trial Case (and how to use it). The ordinary trial case contains about thirty pairs of convex and the same number of concave spherical lenses, ranging from 0.12-D. up to 20-Dioptries;

twenty pairs of convex and the same number of concave cylindrical lenses ranging from 0.12-D, up to 6-Dioptry; at least ten prisms from 1 to 10°; a plain red tinted glass; some shades of smoked glasses; an opaque disc; stenopaic slit; pinhole disc; a ground glass disc; a Maddox rod, or double prism, and a retinoscope; two trial frames, one having three cells to be

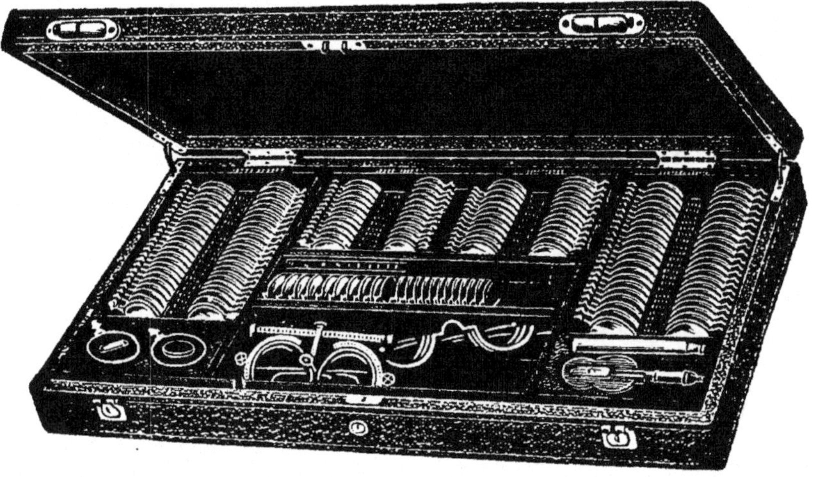

Trial Case.

used in fitting, the other two cells, so that we may allow a patient to wear his correction for a short time and still have one to use. The patient is seated 20 feet from the test card, which must be well illuminated, and shades arranged so that the light will throw no direct rays on the patient's face. Place a small table holding your trial case on the patient's right hand side; seat yourself at the table with your back to the reading chart. Now you will find yourself in a very easy position to change the lenses. The trial frame is placed upon the patient's face and adjusted so that he will look through the center of the lenses, having the frame as near the face as possible. Now you are in a position to begin testing. First place the opaque disc over the left eye, always making it a rule to test the right eye first, as you will find all prescription blanks made out in this way. Now instruct your patient to read the smallest line of letters that he can see with the naked eye. We will say in this case he read the line marked 60. As he is seated 20 feet from the chart, vision with the naked eye is 20/60. You must always remember what the vision with the naked eye is, so that you will be able to judge whether or not the vision is improved with the correction. Now take a plus sphere from the trial case, say plus .50-D., place it before the right eye, asking the patient to again read the smallest type that he can see clearly. Should the patient not read as well, the case may be one of emmetropia, myopia, or astigmatism; but, on the other hand, if he reads just the same as before, or a line better, it is a case of hypermetropia, and we will now proceed to work out a case of each kind.

Hypermetropia. In this case we will say he read line numbered 50. Then his vision will be 20/50. Now we place before this right eye a plus .50 sphere, and if the patient reads the same or a line better, it is surely a case of hypermetropia. Now, as a plus lens will always relax accommodation, and we do not want any eye to accommodate for 20 feet, or farther, we will add more plus in the following manner: take a plus 1 sphere and place it in the second cell of your trial frame, then withdraw the plus

.50. In this way the eye will not be left uncovered; again ask the patient to read, and should he read as well as before we will increase the plus sphere until the smallest line that he reads becomes blurred, then we will know that he has relaxed all the accommodation he had in use; that being the object of the fogging system. Then draw the patient's attention to astigmatic

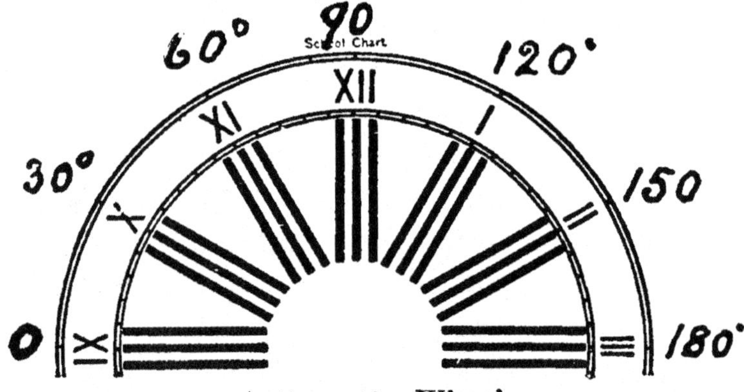

Astigmatic Wheel.

wheel, asking him, "are all the spokes in the wheel equally clear and of the same density?" If there is no astigmatism the patient will see the wheel uniformly. In that case we would ask him to again look at the reading chart, and gradually reduce the strength of the plus sphere, until we find the strongest that will allow the best vision. This will be his correction. On the other hand, had the patient told you that the wheel did not look uniform, but that one or more of the spokes were much darker, it would indicate astigmatism. and we would ask the patient which spoke appeared the most clearly. Now, suppose he says "it is the vertical," or the spoke running from 12 to 6, then as we wish to know if the patient sees the spoke quite clearly, we will ask him to count the lines in the spoke. Should he count the right number we will consider he is seeing it clearly, and to make sure that he is not still accommodating before we correct the astigmatism, we will increase the plus sphere (already in the trial frame) until we just about blur all the spokes in the wheel; then reduce your sphere a quarter D. at a time, at the same time asking the patient to inform you when one of the spokes comes out clearly, and he can count the lines. Whatever plus sphere you have in the frame at this time, place in the cell nearest the eye; or, a better way would be to place a plus sphere of the same strength as the one already in the frame in the cell nearest the eye before removing the one in front. In this way you will move the lens in the frame without exposing the naked eye. It will then be out of the way while using the cylinder. Now take from your trial case the weakest minus cylinder and place it in the trial frame with the axis at right angles to the plainest spoke seen. Should this fail to make the wheel look uniform, increase the strength of your cylinder until you find the weakest that will make the wheel look equal in density in all its spokes. When you have done this, draw the patient's attention to the reading chart, and gradually reduce the strength of your plus sphere while it improves the distant vision. In other words, the strongest plus sphere combined with the weakest minus cylinder that corrected the astigmatism is the patient's correction for constant use.

Myopia. Seat the patient as in the previous

case. Cover the left eye with the opaque disc, ask him to read the smallest type he can with the naked eye, record this vision to compare it with the final correction. Now place a plus .50 in the trial frame, and if the patient is myopic he will say, "I cannot see so well," or in other words, will not be able to read the same line as before. Then draw his attention to the astigmatic wheel and say, "can you see the spokes in the wheel, and do they look equally clear?" If he cannot see any of the spokes clearly enough to count the lines, remove a quarter D. of the plus sphere. If with this he fails to see any of the spokes clearly, remove the other quarter from the trial frame. If none of the spokes are yet clear, begin with the weakest minus sphere and gradually increase same a quarter D. at a time until one or more spokes come up clearly. If they all appear clear at the same time there is no astigmatism, and we turn to the reading chart and give him the weakest minus sphere that will allow him to read the best. This would be his correction. On the other hand, if there is astigmatism, the wheel will not come up equally clear, but some spokes will be plainer than others. The main point is not to increase the minus sphere after one or more spokes appear clearly; for instance, we will say we have on a minus 1 sphere and the patient tells us that he cannot count any lines in any of the spokes as yet. We add to this sphere a minus .25, which will make it minus 1.25, and if he says "Now I can see one spoke clearly," and it runs from 12 to 6, this is the time to begin with the weakest minus cylinder,

placing the axis at right angles to the plain spoke, increasing its strength until you find the weakest that makes the wheel look uniform in density. In this case we will say that it required a minus .75 cylinder, that cylinder combined with the sphere already in the frame will be the correction, which will read as follows·
— 1.25 sph.⊃— .75 cyl. ax. 180.

Triangle (tri′-ang-gl). (L. tres = three + angulus = angle.) A three-sided plane figure.

Trichiasis (trick-i′-a-sis). (Gr. thrix = hair.) That condition where the eyelashes, instead of extending forward, are directed more or less backward, so as to come in contact with the cornea. Trichiasis causes a continual irritation of the eyeball, due to the action of the cilia (eyelashes); there is photophobia, lachrymation, and a constant sense of a foreign body in the eye. The cornea itself suffers considerable injury.

Trichitis (trick-i′-tis). (Gr. thrix = hair + itis = inflammation.) Inflammation of the root of the eyelashes.

Trichosis (tri-ko′-sis). (Gr. thrix = hair.) A disease of the hair. See Trichiasis.

Trichroic (tri-kro′-ik). (Gr. trichroos = three colored.) That which exhibits three different colors in three different positions.

Trichromatic (tri-kro-mat′-ik). (Gr. tri = three + chroma = color.) That which has three colors.

Trigeminus (tri-jem′-in-us). (L. tri = three + geminus = double.) T. **Nerve,** also known as the fifth, or **trifacial,** is the largest cranial nerve.

It has two roots, motor and sensor; it is the sensory nerve of the head and face and motor nerve of the muscles of mastication. It is a branch of this nerve that forms the ophthalmic. (See Nerve.)

Trigonometry (trig-o-nom′-e-try). (Gr. triangle + metry.) That branch of mathematics which treats of the relations of the sides and angles of triangles with the methods of deducing from certain given parts other required parts and also of the general relations which exist between the trigonometrical functions of arcs or angles. Plane trigonometry and spherical trigonometry are those branches of trigonometry in which its principles are applied to plane triangles and spherical.

Triplet. (Gr. tri = three.) A combination of three lenses.

Triplopia. (Gr. triploos = triple + ops = eye.) A visual defect in which three images are seen of the one object.

Trochlea (troch′-le-ah). (L. pulley.) A pulley-shaped part, such as that through which the superior oblique muscle passes.

Trochlearis (troch-le-a′-ris). That which refers to the superior oblique muscle.

Tropom′eter. (Gr. trope = a turning + metron = measure.) An instrument for measuring the movements of the eye.

Tumor (tu′mor). (L. tumere = to swell.) A swelling. A growth of new tissue, differing in structure from the part on which it grows, not the result of inflammation.

Tunic. (L. tunica = coat.) A name given to different membranes, which envelop organs; the eye has three tunics from without inward; first, the sclerotic and cornea; second, choroid, ciliary body and iris combined; third, the retina which is the only one sensitive to light.

Tunica. Same as tunic. **T. adnata,** that portion of the conjunctiva which comes in contact with the eyeball.

Tutam'ina Oculi. (L. tutamen = a protection.) The protecting appendages of the eye, such as the eyelids and lashes.

Tylosis (ty-lo'-sis). (Gr. "knot" + suffix osis = condition.) A thickened, ulcerated condition of the lid margins after ulceration.

Typhlol'ogy. (Gr. typhlos = blind + logia = discourse.) A treatise on blindness.

Typhlo'sis. (Gr. typhlos = blind + osis = condition.) Blindness.

ULCER. (L. ulcus.) An open sore, other than a wound.

Ulceration (ui-ser-a'-shun). Formation of an ulcer.

Umbo (um'-bo). (L. prominence.) The apex, pointed or protuberant part of any substance. When applied to lenses, the extreme elevation of a convex spherical lens, or it may apply to the center of a concave spherical lens.

Umbra. (L. umbra = a shadow.) A shadow.

Undula'tion. (L. unda = wave.) A wave-like motion in any medium.

Un′dulatory Theory. A theory that light, heat and electricity move with a wave-like motion.

Uniaxial (u-ne-ak′-se-al). (L. unus = one.) That which has but one axis.

Unioc′ular. (L. unus = one + oculus = eye.) Only one eye.

Unit (u′-nit). (L. unus = one.) Any standard quantity by the representation and subdivision of which any other quantity of the same kind is measured.

Uremia (u-re′-me-ah). (Gr. "urine" + haima = blood.) Blood poisoning from retained urinary excretions.

Uvaeformis (u-ve-for′-mis). (L. uva = grape + forma = form.) The middle coat of the choroid.

Uvea. (L. uva = grape.) The choroid, ciliary body, and iris together.

Uveal (u′-ve-al). That which refers to the vascular layer of the choroid coat, or the ciliary body and iris.

Uveal Coat. The second tunic or coat of the eyeball.

Uveitis (u-ve-i′-tis). (L. uva = grape + itis = inflammation.) That condition in which the uvea is inflamed. Iritis.

V. Abbreviation for vision.

Vein. (L. venio = I proceed.) The blood-vessels which convey blood toward the heart. They are found wherever there are arteries. Veins have three coats like the arteries, but they are not so thick. The veins draining the eyeball correspond in their arrangement to that of the arteries, the main groups being the retinal, anterior and posterior ciliary veins. The venae vorticosae, collects the blood from the choroid, ciliary body, and the iris and pierces the sclerotic coat near its equator as four large trunks about equal distance from one another, where it is joined by the episcleral veins. The anterior ciliary veins receive the blood from the ciliary muscle and Schlemm's Canal, and after emerging from the sclerotic coat it receives as tributaries the episcleral and vessels from the conjunctiva. These veins upon emerging from the eye unite with the other veins from muscles and tissue to form two main trunks, the superior and inferior ophthalmic, which run along the roof and floor of orbit to terminate in the cavernous sinus. The ophthalmic vein joins the internal angular vein (facial) at the inner and outer angle of the orbit, thus escaping from the orbit in either direction. The inferior ophthalmia vein arises in the veins of the eyelids and lacrimal sac, it also receives the veins from the floor of the orbit and passes out through the sphenoidal fissure.

Velocity. (L. velocitas, from velox = swift.) Quickness of motion.

Venae Vortico'sae ("a whirlpool"). The veins which principally form the external or venous layer of the choroid coat (second tunic) of the eye; so called from their peculiar arrangement, four of them passing out about halfway back.

Ventricle (ven'-tri-kl). A small belly-like cavity, as the two inferior cavities of the heart and various cavities in the brain, of which there are five in number, namely, the two lateral, the third, fourth and fifth. The lateral ventricles are the cavities of each half of the cerebrum. The third ventricle is between the optic thalami at the base of the brain. The fourth V. is the space between the cerebellum and the medulla oblongata. Fifth V. is in the septum lucidum.

Vertex (ver-teks). (L. vertex = top from vertere = to turn.) A turning point; the principal or highest point; top; summit. The point in any figure opposite to, and farthest from, the base; the terminating point of some particular line or lines in a figure or a curve; vertex of a curve is the point in which the axis of the curve intersects it. Vertex of an angle is the point in which the sides of the angle meet. Vertex of a solid or of a surface of revolution is the point in which the axis pierces the surface.

Vertical. Situated at the vertex or highest point. Vertical line is a line perpendicular to the horizon. Vertical plane. A plane passing through the vertex of a cone, and through its axis. (See Vertex.)

Virtual Focus. (L. virtus = power + "the hearth.") An imaginary or negative focus.

Visibility (vis-i-bil'-i-ty). That which has the capacity of being seen.

Vision. (L. videre = to see.) The ability of the organ of sight (the eye) to recognize surrounding objects. **Double v.**, see Diplopia. **Binocular v.**, seeing an object with both eyes at the same time without diplopia. **Monocular v.,** the act of seeing with only one eye.

Visual. Pertaining to vision or sight. **V. Angle,** an angle formed by lines drawn from the extreme edges of an object which cross at the nodal point. **V. Axis,** a line drawn from the macula lutea through the nodal point to the object looked at. **V. Field,** the space containing all objects visible while the eye is in a fixed position. **V. Purple,** purple pigment to be found in the retina, which is bleached by the action of light.

Visual Acuteness. The amount seen by the naked eye if emmetropic; if ametropic, while wearing his correction. The smaller the objects that the eye can distinguish, or the greater the distance at which an object of given size can be seen, the greater is the acuity of vision the eye possesses.

Vitreous (vit′-re-ous). (L. vitrum = glass.) A transparent fluid occupying the posterior and interior four-fifths of the eye.

Vitreous Humor. (L. vitrum = glass + humere = to be moist.) A transparent, colorless, gelatinous mass which fills the posterior cavity. four-fifths of the eye. It somewhat resembles the white of an egg and is surrounded by the hyaloid membrane. See Anatomy. Its index of refraction is 1.33.

Volume (vol′-um). (L. volvere = to roll around.) Solid contents.

Von Graefe's Sign. That condition where the lid fails to move downward with eyeball in exophthalmic goiter.

WALL-EYE. This term has several meanings. It generally refers to white opacities of the cornea or a pale blue iris. Sometimes divergent strabismus.

Wave Theory. The theory that light travels in waves instead of rays. See Light.

Wink. The act of opening and closing the eyelid suddenly.

Winker. See Eyelash.

Worsted Test. The common test employed for color-blindness.

XANTHELASMA (zan-thel-as′-mah). (Gr. xanthos = yellow + elasma = a beaten metal plate.) That condition in which there is a flat tumor of a dirty sulphur-yellow color which projects a little above the skin of the lid. It is found most frequently on the upper and lower lids at the inner angle of the eye.

Xanthocyanopia (zan-tho-cy-an-o′-pi-ah). (Gr. xanthos = yellow + kyanos = blue + ops = eye.) That condition in which there is an inability to perceive red and green colors.

Xanthoma (zan-tho′-mah). (Gr. xanthos = yellow.) A yellowish new growth on the skin.

Xanthophane. (Gr. xanthos = yellow.) A condition in which objects appear yellow.

Xeroma (ze-ro′-mah). (Gr. xeroa = dry.) That condition where the conjunctiva is abnormally dry.

Xerophthalmia (ze-rof-thal'-mi-ah). (Gr. xeros = dry + ophthalmos = eye.) Conjunctivitis with atrophy and no liquid discharge.

Xerosis (ze-ro'-sis). (Gr. xeros = dry.) Abnormal dryness of the eye.

YELLOW SPOT. The macula lutea.

Young-Helmholtz Theory of Color Blindness. (Thomas Young, English physicist, 1773-1829; Herman Ludwig Helmholtz, German physicist. 1821-1894.) The theory that color vision depends on three sets of retinal fibers which correspond to the colors red, violet and green.

ZEISS'S GLANDS. The sebaceous or sweat glands located at the free border of the eyelids.

Zinn's Ligament. (Johann Gottfried Zinn, German anatomist, 1727-1759.) A circular ligament at the optic foramen from which arises the recti muscles of the eye; the ligament itself is attached to the bone and allows the optic nerve to pass through its center.

Zone. (L. girdle.) A girdle or belt.

Zonula. A very small membrane surrounding a body. A small zone.

Zonule of Zinn. The suspensory ligaments of the eye-lens form the Zone of Zinn. It consists of delicate fibers which take their origin from the inner surface of the ciliary body, beginning at the ora serrata. The fibers are in contact with the surface of the ciliary body, but leave it at the apices of the ciliary processes, and, becoming free, divide and pass over to the edge of the lens, thus forming the anterior and posterior suspensory ligaments. These ligaments are attached to the capsule of the lens with which they become fused. The space, triangular in shape, included between the fibers of the zonule or suspensory ligaments and the edge of the lens is called the Canal of Petit.

Just outside of the optic nerve, where it pierces the eyeball, is found a circle of blood-vessels giving a free supply to the optic sheath at this point, and sending branches into the substance of the nerve to supply nutrition. This circle is known as the Circulus of Zinn or sometimes called a Zone of Zinn.

Zonulitis. (L. zonula + Gr. itis.) Inflammation of the zonule of Zinn.

A FEW QUESTIONS WITH THEIR ANSWERS

1. Q. What governs the passage of light through any transparent media?
   A. Density.
2. Q. On what does the visual angle depend for its existence?
   A. The size and distance of the object.
3. Q. What three laws accompany refraction?
   A. Reflection, absorption and dispersion.
4. Q. In what three ways can an incident ray be disposed of?
   A. Reflected, absorbed or refracted.
5. Q. What three laws must be brought into play in order to obtain distinct binocular vision at various distances?
   A. Refraction, accommodation and convergence.
6. Q. Why is it necessary for the aqueous humor to be thinner than the vitreous humor and yet have the same density?
   . To allow freedom of movement to the iris.
7. A. Why is accommodation and convergence so closely associated?
   A. Because they are both operated by the same nerve.
8. Q. What lens represents the focal strength of the dioptric system of the eye?
   A. From 62 to 65-D. plus.
9. Q. When is a lens periscopic?
   A. When it is minus on one side and plus on the other.
10. Q. What are objective and subjective symptoms?
    A. Objective symptoms are what the operator detects without questioning the patient.

Subjective symptoms are those described by the patient.

11. Q. Why does Astigmatism in one eye sometimes cause convergent Strabismus?

A. In order to prevent the eye with the blurred vision interfering with the vision of the good eye the patient learns to turn it toward the nose.

12. Q. Why do we add and subtract from retinoscopic findings?

A. To place the patient's far point at 20 feet.

13. Q. Why is the concave retinoscope superior to the plane?

A. Because a concave retinoscope combined with a plus 20-D. lens can be used as an ophthalmoscope.

14. Q. What lens can be combined with plus 2 sphere combined with plus 1 cylinder, axis 90, that will increase the cylinder and decrease the sphere?

A. Any minus cylinder under 2 dioptries with its axis at 180.

15. Q. What is false myopia, and how is it produced?

A. A spasm of accommodation in emmetropia will cause the eye to appear myopic, and is brought about by continual strain at close work, exophoria or hyperopia.

16. Q. Is an Ophthalmic lens used to improve vision?

A. No. The lens is to refract the light only, which sometimes brings the focus nearer to the retina resulting in better vision.

17. Q. Could Strabismus exist and the eyes be Orthophoric?

A. Yes. The deviation may be due to Hypermetropia or abnormal vision in one eye only.

18. Q. Can an eye be parallel with its fellow while wearing a single prism?

A. Not unless the Macula Lutea is displaced.

19. Q. Do prisms make eyes parallel in Heterophoria?

A. No. They allow them to deviate from parallelism.

20. Q. Name three things that would interfere with the action of accommodation.

A. Paralysis, hardening of the lens, loss of Crystalline Lens.

21. Q. What three things determine the focal length of a lens?

A. Curvature, density and distance of object.

22. Q. What is a prism diopter?

A. The power of a prism to deviate a ray of light one centimeter in a meter.

23. Q. What are the constituents of Crown Glass?

A. Crown glass is so called from the trademark by the early manufacturers. It is composed of lime, sand and soda.

24. Q. What is the composition of Flint Glass?

A. Somewhat like Crown only less sand and lead is used to increase its density.

25. Q. Which is the harder, flint or crown glass?

A. Crown glass is about 20 per cent harder, because it has no lead and more sand.

26. Q. Which has the most dispersive power, flint or crown?

A Flint has the most dispersive power, because of its higher index of refraction.

27. Q. Transpose the following Rx.:
—1. cyl. ax. 20
—2. sph.+2.50 cyl. ax. 180
+ .50 sph.+1 cyl. ax. 60

A. —1 sph. + 1 cyl. ax. 110
+.50 sph. —2.50 cyl. ax. 90
+ 1.50 sph. —1 cyl.ax.150

28. Q. What errors of refraction will the above prescriptions correct?

A. (a) Simple myopic astigmatism. (b) Mixed astigmatism. (c) Compound hypermetropic.

29. Q. What vision does an emmetropic eye always have?

A. Emmetropia has no reference whatever to vision, but we expect normal.

30. Q. Does a hypermetropic eye ever have normal vision?

A. Yes. Provided he is able to overcome his error by accommodating.

31. Q. Name four reasons why vision would not be normal?

A. Cataract. Focus off the retina. Amblyopia. Lucoma.

32. Q. If an object is 80 inches from a +3 sphere where will the image be?

A. 16 inches.

33. Q. If an eye looking 20 feet away sees as well without as with a +2 sph. should it be worn? Give reason for your answer.

A. Yes. It shows relaxation from accommodation.

34. Q. In correcting astigmatism where would you place the axis of the cylinder, if the spoke from 11 to 5 was seen the blackest?

A. 150 degrees.

35. Q. What error of refraction would you cause and would it be with or against the rule if you placed a +1 cyl. ax. 180 before an emmetropic eye?

A. Simple Myopic astigmatism. With the rule.

36. Q. Upon what is the deviating power of a prism dependent?

A. The power of prism is dependent upon its angle and its index of refraction.

37. Q. What is a compound lens?

A. A lens made with a spherical and cylindrical effect, also known as a sphero-cylinder.

38. Q. What is the kind of glass from which ordinary spectacle lenses are made? What is its index?

A. It is Crown Glass. Index varies from 1.50 to 1.58.

39. Q. What does myosis with photophobia in normal light indicate?

A. Myosis in normal light with photophobia indicates a serious disease, as a rule.

40. Q. What is the first thing to do in making an examination?

A. Examine for diseased conditions before correcting errors of refraction.

41. Q. How would you prove a pair of eyes has exophoria that requires a prismatic correction for constant wear?

A. If after the patient has worn his distant correction for the ametropia a few weeks the muscle trouble remains, I would figure to correct the trouble. With the maddox rod horizontal before the right eye and the red glass before the left, the patient will see the red light to the right of streak, and it would require prisms base in to bring them together, proving exophoria.

Points one should be familiar with before attempting a State Examination:

1. Mechanical parts of frames and guards for mounting lenses; making face measurements for same; truing up bent frames and guards; adjusting same to different persons.

2. The common shapes and forms and dioptric values of lenses of different kinds; submitting ten different kinds to applicants for determination of these qualities.

3. Practical fitting with trial case, a test of the applicant's practical ability to go through these tests and accurately fit different classes of cases with lenses.

4. Shadow testing, with or without stand instrument; the actual doing of this work and determining the error of refraction by the method. The mirror or instrument preferred may be used.

5. Muscle testing, and the use of muscle testing devices; a test of the applicant's ability to make these tests and draw correct conclusions from them and their showings.

6. The proper use of different optical instruments used to measure the refraction of the eyes or any surface, or the power of the muscles of the eyes.

7. Questions on the anatomy and physiology of the eyes, including muscles, nerves, tissues and their functions.

8. Questions on refraction of lenses, transpositiou, conjugate foci, image forming, and the

GROUP ONE.

1. Name the extrinsic and intrinsic muscles of the eye.

2. Describe the Iris, the muscles that control its movements and the mechanism that causes the pupil to contract and expand, what causes the movements and how they assist in procuring clear vision.

3. Describe the humors of the eye and why it is desirable that a certain humor be more fluid than the others.

4. Describe the Choroid and its functions.

5. Describe the Crystalline Lens. What holds it in position? What controls and alters its shape? What is its index of refraction? What is its strength in diopters and how does it assist in procuring good vision?

6. Describe the Optic Nerve and Retina and their functions.

7. Is the human eye chromatic or achromatic? How can it be proven?

8. Describe the Conjunctiva and its office.

9. Which part of the eye has the greatest refracting power? Why?

10. Describe the eyelids, eyebrows, eyelashes and the office of each.

11. Describe the Lachrymal Gland and its appendages and their office.

12. In which part of the brain are the optic centers situated?

13. Describe the principal veins and arteries of the eye.

14. From what source does the Cornea receive its supply of nourishment?

15. What is meant by abduction? Adduction?

Sursumduction? Torsion? Name the muscles that produce them.

16. Define visual angle; minimum visual angle; visual acuity.

17. What is Pterygium? Trachoma? Nystagmus? Keratitis?

18. Explain the Helmholts and Tscherning theories of accommodation. Which, in your opinion, is more reasonable? Why?

19. What do you understand by Co-ordinate movements of the eye?

20. Make a diagram of an eye, showing the principal parts, briefly describing the functions of each part.

GROUP TWO.

1. Explain what is meant by the Conjugate movements of the eyes.

2. What is the refractive condition and amplitude of accommodation of a person whose P. R. with a plus 5 D. Sphere is 50 C. M. and whose P. P. with a plus 2 D. Sphere is 10 C. M.?

3. What is the usual size of the ocular pupil in uncorrected ametropes who do not complain of asthenopia?

4. A person has a fixed deviation outward of one eye of 4 meter angles. What prism will cause the images to fuse at 50 C. M.? Where would you place the base of the prism?

5. Describe fully your method of measuring the relative strength of the extrinsic muscles of the eyes.

6. Give the rule for the prism diopral decentration of lenses, and work the following problem by it: A pair of plus 3 D. S. are decentered 5 mm. inward. How much prism value results?

7. Give the rule for transposing plus spheres combined with plus Cyl's to plus Sph. combined with minus Cyl's and work the following by it (leaving your figures on the paper): (a) Plus 1.75 D. S. combined with plus .75 D. C., axis 15. (b) Plus 1.25 D. S. combined with plus 1.50 D. Cyl. axis 135.

8. Give the rule for transposing cross Cyl's into Sphero Cyl, and work the following by it, leaving your figures: (a) Plus 1 D. C. axis 90 combined with a plus 2 D. C. axis 180. (b) Plus 1.25 D. C. axis 90 combined with a minus .50 D. C. axis 180.

9. In Retinoscopy what is the nature of the refraction of the eye under observation when one point of reversal is found at 20 inches and another at 40 inches?

10. A coin one inch in diameter is held 12 inches from a lens, the principal focus of which is 4 inches. Where will the image be formed?

11. Explain the relation existing between accommodation and convergence.

12. Explain the principles governing the use of the Ophthalmometer and its practical working. To what extent can it be relied upon as a means of estimating refractive errors?

13. A child of 10 years who has never worn glasses is found to have 4 D. of hyperopia and esophoria of 5 degrees. Would you order prisms or decenter the lenses in this case?

14. A certain prescription reads plus 1 D. S. combined with plus 1.25 D. C. axis 90. Give the dioptric value of the surfaces of the lens.

15. Define Anisometropia; Asthenopia; Amblyopia; Orthophoria.

16. With the static method and the retinoscope at 40 inches the shadow is neutral with a plus 2

Sphere before the observed eye. What two points are conjugate and what is correction for that eye?

17. In static skiametry with plane mirror at 26 inches and a plus 1.50 D. S. in rear cell of trial frame no movement is found in the horizontal meridian and a plus .75 Sph. neutralizes motion in the vertical meridian. What direction had the shadow before the plus .75 D. S. was used? What kind and amount of ametropia was indicated?

18. A boy of 14 years has vision equal to 20-15 without glasses. Would you accept that as proof of emmetropia? Give your reasons.

19. If double concave lens could be made of air with its radius of curvature 20 inches for each side, what would be the nature of its refraction if it were immersed in water and what would be its action on parallel rays of light?

20. If a person 15 years of age should come to you to have a complete examination of the eyes, how would you proceed? What questions would you ask and how would you proceed to make the examination, together with the record you would make? Give full particulars just as if you had such a case.

GROUP THREE.

1. What is light? How does it travel from a luminous point? What three things may happen to it when it comes in contact with another medium?

2. What is reflection? Give the law governing incident and reflected rays. If a ray is incident at an angle of 30 degrees, at what angle will it be reflected?

3. The radius of curvature of a concave mirror is 16 inches. What will be its principal focus?

If a candle is placed 12 inches in front of it, at what distance will its image be formed?

4. What is refraction? Make a diagram showing the path of a ray of light as it strikes the surface of a plate of glass, as it passes through, and after it emerges. What are the three parts of the ray called?

5. What is meant by the term "surface power of a lens"? A certain lens has a surface power of plus 6.25 D. S. on one side and on the other minus 1.25 D. S. What curvatures must a cement segment have to make the reading portion plus 6 D. when the index of refraction is the same in both parts of the lens?

6. Make diagram showing three double convex lenses and the course of parallel rays of light through them, the principal focus, the secondary focus, and one pair of conjugate foci.

7. A plus 2.50 D. S. is held 30 inches from a candle. Where will the image be formed?

8. The light from a candle passes through a plus 3 D. S. at 50 C. M. Where should a screen be placed to get the best defined image? Show your method of working this problem.

9. A certain lens focuses parallel rays of light at 20 C. M. What lens must be placed in front of it to send the focus back to one meter?

10. What is meant by "the power of a prism"? If you had two weak prisms and desired to know their powers within one per cent, what method would you employ and how would you proceed to carry out the measurement?

11. Describe how bifocal lenses should be adjusted and the position which the reading portion should occupy.

12. A plus 1.50 D. plano Cyl is placed in contact with a similar one of plus .75 D.; axis right angles. What is the resultant combination?

13. If you desired to find the power of a double concave lens and had no lens measure or neutralizing lenses, how would you proceed if the index of refraction of the glass was 1.50?

14. Give the rule for transposing plus spheres combined with minus Cyls and work the following by it: Plus 3 D. S. combined with minus 1.25 D. C. axis 45. Plus 2 D. S. combined with minus 2.50 D. C. axis 65.

15. Transpose the following:

Plus 3.50 D. S. plus 1.25 D. C. axis 135.
Plus 2 D. S. plus D. C. axis 65.
Minus 3 D. C. minus 2 D. C. axis 75.
Minus 2 D. S. minus 3.25 D. C. axis 155.
Plus 2.25 D. C. axis 90 plus 150 D. C. axis 180.
Minus 2.25 D. S. plus 175 D. C. axis 135.

16. What error will the following prescription correct: R. plus 1 D. S. minus .75 D. C. axis 180, combined with 2 degree prism base up.

17. Describe Glaucoma. Trachoma. Pterygium. Cataract. Granulated lids.

18. What is Ophthalmia Neonatorum? What simple method will prevent it?

19. Describe the fogging method of measuring refraction and your everyday manner of applying it.

20. A young person complains of Asthenopia and the third trial case shows Myopia of 1.25 D. The Retinoscope shows 1.50 of Hyperopia. What would you give in the way of lenses? Why would you give them?